Undergraduate Topics in Computer Science

'Undergraduate Topics in Computer Science' (UTiCS) delivers high-quality instructional content for undergraduates studying in all areas of computing and information science. From core foundational and theoretical material to final-year topics and applications, UTiCS books take a fresh, concise, and modern approach and are ideal for self-study or for a one- or two-semester course. The texts are all authored by established experts in their fields, reviewed by an international advisory board, and contain numerous examples and problems, many of which include fully worked solutions.

The UTiCS concept relies on high-quality, concise books in softback format, and generally a maximum of 275–300 pages. For undergraduate textbooks that are likely to be longer, more expository, Springer continues to offer the highly regarded Texts in Computer Science series, to which we refer potential authors.

More information about this series at http://www.springer.com/series/7592

K. Erciyes

Discrete Mathematics and Graph Theory

A Concise Study Companion and Guide

 Springer

K. Erciyes
Department of Computer Engineering
Üsküdar University
Üsküdar, Istanbul, Turkey

ISSN 1863-7310   ISSN 2197-1781   (electronic)
Undergraduate Topics in Computer Science
ISBN 978-3-030-61114-9   ISBN 978-3-030-61115-6   (eBook)
https://doi.org/10.1007/978-3-030-61115-6

This Springer imprint is published by the registered company Springer Nature Switzerland AG
The registered company address is: Gewerbestrasse 11, 6330 Cham, Switzerland

To all Math lovers

Preface

Discrete Mathematics is a branch of mathematics that studies structures which take distinct values as opposed to continuous mathematics branches such as calculus and analysis. Study of discrete mathematics is one of the first courses of curriculums in various disciplines such as Computer Science, Mathematics and various engineering branches. There are many books on discrete mathematics and it will be worthwhile to justify the need for another book on discrete mathematics.

Graphs are key data structures used to represent various networks, chemical structures, games, etc. Graph theory has gone through an unprecedented growth in the last few decades both in terms of theory and implementations, hence we believe it deserves a thorough treatment in any discrete mathematics book which probably is not found adequately in any of contemporary book of discrete mathematics. This is one of the main reasons of writing this book.

Secondly, many contemporary books on the topic elaborate on basic concepts vigorously with numerous examples and exercises which are appropriate for an undergraduate course. However, we think there is a need for a dense and thorough treatment of discrete mathematics and graph theory possibly for the non-Computer Science majors of various other disciplines. Also, this book can also be used as a reference/second textbook in a Computer Science curriculum for quick reference.

Lastly, we will be using an algorithmic approach for some discrete mathematics and graph problems to reinforce learning the topics and show how to implement these concepts in real applications.

With this in mind, this book is intended as a reference book for a one-semester course in the junior level of Computer Science, Mathematics and senior/graduate levels of various engineering disciplines. We have a web page to provide any supporting material and possible errata at:

http://akademik.ube.ege.edu.tr/erciyes/DMGT.

Part of the material in the book was used for the first year Discrete Mathematics course at Üsküdar University Computer Engineering Department. I would like to thank students of this course for their valuable feedback, especially for some concepts that did not seem so obvious which resulted in further elaboration of these topics in the book. As usual, my thanks go to Springer senior editor Wayne Wheeler and associate editor Simon Rees for their support during the writing of the book.

Üsküdar, Turkey   K. Erciyes

Contents

Fundamentals of Discrete Mathematics

Part I

Logic

Logic deals with laws of thought, and mathematical logic and proofs are the basic methods of reasoning mathematics. The laws of logic help us to understand the meaning of statements. Statements come in various forms, for example "7 is a prime number" is a statement which we can deduce to be true. Sometimes a statement can be of the form "if n is an even integer, we can write $n = 2m$ for some integer m" which is a true compound statement consisting of two simple statements. The first simple statement is "n is an even integer" and the second one is "$n = 2m$". We claim the second statement is true when the first one is true. Statements in the form "if ...then..." form the basis of reasoning about proofs. Yet, determining the truth value of a statement such as "My name is Melisa" depends on the person who says it. We will investigate statements, how to determine their truth values, and the laws that aid to reason about complex statements in this chapter.

1.1 Propositional Logic

A *proposition* or a *statement* is a declaration that is either correct or incorrect. Some examples of propositions are:

1. A prime number cannot be factorized.
2. 12 is an odd integer.
3. It is sunny today.
4. $3 + 4 = 7$.
5. For any integer n, n^2 is a positive integer.

© Springer Nature Switzerland AG 2021
K. Erciyes, *Discrete Mathematics and Graph Theory*, Undergraduate Topics
in Computer Science, https://doi.org/10.1007/978-3-030-61115-6_1

Table 1.1 Truth table for negation

p	$\neg p$
T	F
F	T

However, a statement such as "How old are you?" is not a proposition as we cannot assign a truth value to it. A proposition may be *true* or *false* at all times, and sometimes its *truth value* depends on when or by whom it is stated. For example, although propositions 1, 4 and 5 are true, and 2 is false; the value of proposition 3 depends on the day it is stated. We will use a lowercase letter to represent a statement and "T" for the true value and "F" for the false value of a proposition.

Negation
The negation of a proposition is a proposition with the inverse value of the statement. This is accomplished by inserting a "not" or "it is not the case" in a proposition. The negation of a proposition p is shown by $\neg p$ or $\backsim p$ or $\overline{p}$; we will adopt $\neg p$ throughout the book. For example, if p = "It is rainy today", then $\neg p$ = "It is not rainy today". We could also state $\neg p$ = "It is not the case it is rainy today". A *truth table* lists all of the possible truth values of a proposition. The truth table for the negation is shown in Table 1.1.

1.1.1 Compound Propositions

A proposition may consist of more than one statement joined by *logical connectives*. For example, p = "16 is an even number" *and* "16 > 4". Here, p contains two *component* propositions joined by the word *and*, and its truth value depends on the truth values of these simple component propositions. In this example, p is true since both of its components are true. Two commonly used connectives in a compound proposition are *and* and *or*.

Conjunction
The *conjunction* of two simple propositions is the proposition obtained by inserting *and* which is commonly shown by the logical symbol $\wedge$ between them. Thus, the conjunction of two propositions p and q is denoted by $p \wedge q$. The conjunction is true when both p and q are true and false otherwise. Truth table for conjunction is shown in Table 1.2.

Disjunction
The *disjunction* of two simple propositions is the proposition obtained by inserting "or" or the logical symbol $\vee$ between them. Hence, disjunction of two propositions p and q is denoted by $p \vee q$. It is false when both p and q are false and true otherwise. For example, "6 is a prime number *or* 4 > 9" is a compound proposition consisting of two simple statements that are both false and hence the compound proposition is

Table 1.2 Truth table for conjunction

p	q	$p \wedge q$
F	F	F
F	T	F
T	F	F
T	T	T

Table 1.3 Truth table for disjunction

p	q	$p \vee q$
F	F	F
F	T	T
T	F	T
T	T	T

Table 1.4 Truth table for exclusive or

p	q	$p \oplus q$
F	F	F
F	T	T
T	F	T
T	T	F

false. Truth table for disjunction is shown in Table 1.3 where we can see the truth value of such a proposition is true when at least one of its components yields a true value. A variation of disjunction is the *exclusive-or* of two propositions. This is true when only one of the propositions p and q is true and is false otherwise as shown in Table 1.4. Basically, exclusive-or is used for cases when it is required that only one of the statements that make up the compound statement needs to be true. For example, a student may take one of two similar courses but not both.

1.1.2 Conditional Statements

The *conditional statement* or an *implication*, $p \rightarrow q$, is read as "if p then q" and is false when p is true and q is false, and true otherwise. The proposition p is commonly called the *hypothesis* or *premise*, and q is called the *conclusion* or the *consequence*. The conditional statement can also be stated as follows.

- "p implies q"
- "p only if q"

Table 1.5 Truth table for $p \rightarrow q$

p	q	$p \rightarrow q$
F	F	T
F	T	T
T	F	F
T	T	T

- "p is (a) sufficient (condition) for q"
- "q is (a) necessary (condition) for p"

Example 1.1.1 Let the proposition p be "I study hard" and q be "I pass the exam". The conditional statement would then be "if I study hard, I pass the exam". Note that *not* studying hard does not imply failing the exam, we are only saying studying hard implies passing the exam. Studying hard and failing the exam as a consequence will not happen and therefore is a false condition. Also, if I do not study hard, I may and probably will not pass the exam which is again a legitimate and therefore a true status.   □

The truth table for the implication is shown in Table 1.5 which summarizes the consequences of the possible input statement combinations. It can be seen that the proposition "if p then q" is false only when p is true and q is false since this can not happen based on our assumption that if premise is true, the consequence should also be true. Note that a false proposition implying a true proposition yields an overall true value for the conditional statement. This is logical as the consequence is true as in the conditional "If snow is green then sun is yellow". In other words, we do not care what the truth value of the premise is as long as the conclusion is true. A false premise implying a false conclusion also results in a true value for the conditional as shown in Table 1.5 as in the example "If earth is flat then it can not rotate". We know earth is not flat and also it does rotate. A false premise implying a false conclusion may seem more illogical than this example, "If iron is soft then water is hard" is such a conditional proposition which results in a total true statement. We know iron is not soft, thus, it is not the case that water is hard. Hence, a false premise may imply any conclusion that is false to result in a true conditional value.

Definition 1.1 Given a conditional statement r of the form $p \rightarrow q$,

- The *converse* of r is $q \rightarrow p$
- Its *inverse* is $\neg p \rightarrow \neg q$
- Its *contrapositive* is $\neg q \rightarrow \neg p$

The truth table for the converse, inverse and the contrapositive of a conditional proposition are depicted in Table 1.6. We can see from this truth table that the contra-

Table 1.6 Truth table for converse, inverse and contrapositive of a conditional proposition

p	q	$\neg p$	$\neg q$	$p \rightarrow q$	$q \rightarrow p$	$\neg p \rightarrow \neg q$	$\neg q \rightarrow \neg p$
F	F	T	1	T	T	T	T
F	T	T	F	T	F	F	T
T	F	F	T	F	T	T	F
T	T	F	F	T	T	T	T

positive of a conditional statement is equivalent to itself and this fact is very useful when proving theorems as we shall see. Furthermore, the converse of a conditional proposition is always equivalent to its inverse as shown.

Let us illustrate these concepts by the previous example; let proposition p be "I work hard" and q be "I pass my Math exam" as before. The conditional statement r formed using p and q would then be "If I work hard, I pass my Math exam". The contrapositive of r is "If I do (did) not pass my Math exam, I do (did) not work hard". More informally, this means if I have not passed my Math exam, then I have not worked hard. This is true because if I had worked hard, I would have passed the exam based on the conditional proposition r which shows r and its contrapositive are equivalent. The converse of r then would be "If I pass my Math exam, I (have) work(ed) hard". This is not true since I could have passed my exam even if I had not worked hard for it. Essentially, the conditional r does not specify this case. Its inverse is "If I do not work hard, I will not pass my Math exam" which provides basically the same reasoning as the converse of r since I could have passed the exam without working hard as in the converse.

1.1.3 Biconditional Statements

In some cases, we require the premise p to be both a necessary and a sufficient condition for the conclusion q. Note that this is different than the conditional since we need the premise p to be a necessary condition this time. We use the phrase *if and only if* to connect the two component propositions this time saying "p if and only if q". Returning to our previous example which had p ="I work hard" and q ="I pass the exam" now becomes "I study hard if and only if I pass the exam" in biconditional form. What we are saying this time is that there is no way to pass the exam without working hard. Note that passing the exam without working hard was a possibility in the conditional.

The truth table for the biconditional $p \leftrightarrow q$ is shown in Table 1.7 and based on our above discussion, it has true values when both component propositions have the same value whether true or false. Here, we can see that a false statement implying a false statement yields a true value of the biconditional. Later, we will see that a biconditional statement is equivalent to conjunction of two conditionals, one from each direction.

Table 1.7 Truth table for $p \leftrightarrow q$

p	q	$p \leftrightarrow q$
F	F	T
F	T	F
T	F	F
T	T	T

Example 1.1.2 Let the proposition p be "it is sunny" and q be "I go to park". Write the following propositions using p and q:

1. If it is sunny, I go to park.
2. If I do not go to park, it is not sunny.
3. I go to park if and only if it is sunny.

Solution:

1. $p \rightarrow q$
2. $\neg q \rightarrow \neg p$
3. $q \leftrightarrow q$

□

1.1.4   Tautologies and Contradictions

Some compound propositions are always true or false irrelevant of the truth values of the simple statements that make them. A *tautology* is a compound statement that is always true independent of the truth values of is component statements. Similarly, a *contradiction* is always false immaterial of the truth value of its components. The truth table for an example tautology and an example contradiction are shown in Tables 1.8 and 1.9. We have the truth value of a proposition *or* its negation which is always true, and the truth value of a proposition *and* its negation which is always *false*.

Table 1.8 Truth table for a tautology

p	$\neg p$	$p \vee \neg p$
F	T	T
T	F	T

Table 1.9 Truth table for a contradiction

p	$\neg p$	$p \wedge \neg p$
F	T	F
T	F	F

Table 1.10 Truth table for $(p \rightarrow \neg q) \vee (\neg r \rightarrow p)$

p	q	r	$\neg q$	$\neg r$	$p \rightarrow \neg q$	$\neg r \rightarrow q$	$(p \rightarrow \neg q) \vee (\neg r \rightarrow q)$
F	F	F	T	T	T	F	T
F	F	T	T	F	T	T	T
F	T	F	F	T	T	T	T
F	T	T	F	F	T	T	T
T	F	F	T	T	T	F	T
T	F	T	1	F	T	T	T
T	T	F	F	T	F	T	T
T	T	T	F	F	F	T	T

Table 1.11 Truth table for $(\neg p \wedge q) \wedge (p \vee \neg q)$

p	q	$\neg p$	$\neg q$	$\neg p \wedge q$	$p \vee \neg q$	$(\neg p \wedge q) \wedge (p \vee \neg q)$
F	F	T	T	F	T	F
F	T	T	F	T	F	F
T	F	F	T	F	T	F
T	T	F	F	F	T	F

Example 1.1.3 Show that $(p \rightarrow \neg q) \vee (\neg r \rightarrow p)$ is a tautology.

Solution: We form the truth table for the propositions p, q and r, this time with 8 rows to include all of the possible 2^3 input values for 3 variables as in Table 1.10. The required compound proposition is in the last column of this table and we find it has all true values and therefore this compound statement is a tautology. $\square$

Example 1.1.4 Find whether $(\neg p \wedge q) \wedge (p \vee \neg q)$ is a contradiction.

Solution: We form the truth table for the propositions p, q, their negations and the component propositions as shown in Table 1.11. We form the required compound proposition in the last column of this table and find it has all false values and therefore this compound statement is a contradiction. $\square$

1.1.5 Equivalences

Two propositions are considered *equivalent* if they have the same truth table values for all their simple component proposition truth value combinations. We show the equivalence by "≡" sign, that is, $p \equiv q$ if p and q are equivalent to each other. Table 1.12 shows that the biconditional $p \leftrightarrow q$ is equivalent to $(p \rightarrow q) \wedge (q \rightarrow p)$ as shown in the last two columns. This means the assessment of the truth value of a biconditional statement needs to be done by assessing the truth value of both of the two conditional statements $(p \rightarrow q)$ and $(q \rightarrow p)$.

Example 1.1.5 Show that $p \rightarrow q \equiv \neg p \vee q$ and $\neg(p \rightarrow q) \equiv p \wedge \neg q$
Solution: Let us form the truth table for both propositions against all possible values of the simple statements as shown in Table 1.13. The corresponding columns (4 and 5) for the two compound propositions are equal as can be seen. For the second equivalence, columns 7 and 8 are equal.   □

These two equivalences play an important role in proving the conditional statements. For the first equivalence $p \rightarrow q \equiv \neg p \vee q$, let p be "You do not need me now" and q be "I will go out". The implication is "If you do not need me now, I will go out". The equivalent statement based on the above example is "Either you need me now or I will go out". For the second one $\neg(p \rightarrow q) \equiv p \wedge \neg q$, we can state "If it is not the case that you do not need me now then I will go out" is equivalent to "You do not need me now and I will not go out".

We now have the following observation. When two propositions p and q are equivalent, they have the same truth table values as stated. On the other hand, the

Table 1.12 Truth table of the equivalence of a biconditional

p	q	$p \rightarrow q$	$q \rightarrow p$	$(p \rightarrow q) \wedge (q \rightarrow p)$	$p \leftrightarrow q$
F	F	T	T	T	T
F	T	T	F	F	F
T	F	F	T	F	F
T	T	T	T	T	T

Table 1.13 Truth table for Example 1.1.5

p	q	$\neg p$	$p \rightarrow q$	$\neg p \vee q$	$\neg q$	$p \wedge \neg q$	$\neg(p \rightarrow q)$
F	F	T	T	T	T	F	F
F	T	T	T	T	F	F	0
T	F	F	F	F	T	T	T
T	T	F	T	T	F	F	0

Table 1.14 Truth table for Example 1.1.6

p	q	$\neg p$	$\neg q$	$p \wedge q$	$a: \neg(p \wedge q)$	$b: \neg p \vee \neg q$	$a \leftrightarrow b$
F	F	T	T	F	T	T	T
F	T	T	F	F	T	T	T
T	F	F	T	F	T	T	T
T	T	F	F	T	F	F	T

Table 1.15 Truth table for Example 1.1.7

p	q	$\neg p$	$\neg q$	$p \vee q$	$a: \neg(p \vee q)$	$b: \neg p \wedge \neg q$	$a \leftrightarrow b$
F	F	T	T	F	T	T	T
F	T	T	F	T	F	F	T
T	F	F	T	T	F	F	T
T	T	F	F	T	F	F	T

biconditional statement is true whenever its components have the same value. We can therefore conclude $p \equiv q$ if and only if $p \leftrightarrow q$ is a tautology.

Example 1.1.6 Let us check whether $\neg(p \wedge q) \leftrightarrow \neg p \vee \neg q$ is a tautology. We form the truth table shown in Table 1.14 and the last column in this table corresponding to the required statement has all true values resulting in a a tautology. □

Example 1.1.7 Let us show that $\neg(p \vee q) \leftrightarrow \neg p \wedge \neg q$ is a tautology. The truth table shown in Table 1.15 has all true values in the last column for the compound proposition meaning this statement is a tautology. □

We have in fact proven two important laws in logic called *De Morgan's Laws* in the last two examples.

Example 1.1.8 Show that $p \oplus q \equiv (p \vee q) \wedge \neg(p \wedge q)$.

Solution: The truth table shown in Table 1.16 has all equal values for rows in the last two columns for the left and right sides of the equivalence meaning they are equal. □

Table 1.16 Truth table for Example 1.1.8

p	q	$\neg p$	$p \vee q$	$p \wedge q$	$\neg(p \wedge q)$	$(p \vee q) \wedge$ $\neg(p \wedge q)$	$p \oplus q$
F	F	T	F	F	T	F	F
F	T	T	T	F	T	T	T
T	F	F	T	F	T	T	T
T	T	F	T	T	F	F	F

1.1.6 Laws of Logic

We are now ready to state the laws of logic shown in Table 1.16 many of which are basically common sense. For example, the idempotent law says $p \vee p$ is p which states that disjunction of a proposition by itself is the proposition. However, some of these laws such as De Morgan's Laws may not be easy to see at once.

	Logical Equivalence Laws
Identity Laws:	$p \vee F \equiv p$
	$p \vee T \equiv T$
	$p \wedge F \equiv F$
	$p \wedge T \equiv p$
Negation Laws:	$p \vee \neg p \equiv T$
	$p \wedge \neg p \equiv F$
	$\neg\neg p \equiv p$
Idempotent Laws:	$p \vee p \equiv p$
	$p \wedge p \equiv p$
Commutative Laws:	$p \vee q \equiv q \vee p$
	$p \wedge q \equiv q \wedge p$
Associative Laws:	$(p \wedge q) \wedge r \equiv p \wedge (q \wedge r)$
	$(p \vee q) \vee r \equiv p \vee (q \vee r)$
Distributive Laws:	$p \vee (q \wedge r) \equiv (p \vee q) \wedge (p \vee r)$
	$p \wedge (q \vee r) \equiv (p \wedge q) \vee (p \wedge r)$
Absorption Laws:	$p \vee (p \wedge q) \equiv p$
	$p \wedge (p \vee q) \equiv p$
De Morgan"s Laws:	$\neg(p \vee q) \equiv \neg p \wedge \neg q$
	$\neg(p \wedge q) \equiv \neg p \vee \neg q$

We now have a second method to show the equivalence of two propositions. We can use these laws of equivalences rather than constructing a truth table and checking each row of the logical statements to be the same for all input combinations. This new method is convenient since for n simple input propositions, the truth table would contain 2^n rows which may be difficult to realize when $n > 3$.

Example 1.1.9 We want to show that $\neg(p \lor q) \lor (\neg p \land q) \equiv \neg p$. Let us use the logic laws for the left side of this equivalence:

$$\neg(p \lor q) \lor (\neg p \land q) \equiv (\neg p \land \neg q) \lor (\neg p \land q) \qquad \text{De Morgan}$$
$$\equiv \neg p \land (\neg q \lor q) \qquad \text{Distributive Law}$$
$$\equiv \neg p \land T \qquad \text{Negation Law}$$
$$\equiv \neg p \qquad \text{Identity Law}$$

$\square$

Example 1.1.10 We want to show that $p \land (\neg p \lor q) \equiv p \land q$. Using the laws for the left side of this equivalence yields,

$$p \land (\neg p \lor q) \equiv (p \land \neg p) \lor (p \land q) \qquad \text{Distributive Law}$$
$$\equiv F \lor (p \land q) \qquad \text{Negation Law}$$
$$\equiv p \land q \qquad \text{Identity Law}$$

$\square$

Example 1.1.11 Let us prove the equivalence:

$$(p \land q) \to r \equiv (\neg p \lor \neg q) \lor r$$

We can start by what we know from Example 1.1.5:

$$a \to b \equiv \neg a \lor b$$

and implementing this property in the above example yields:

$$(p \land q) \to r \equiv \neg(p \land q) \lor r$$
$$\equiv (\neg p \lor \neg q) \lor r \qquad \text{De Morgan}$$

$\square$

Example 1.1.12 Prove that $(p \land q) \to p$ is a tautology.
Solution: Let us use laws of logic to prove this statement:

$$(p \land q) \to p \leftrightarrow \neg(p \land q) \lor p$$
$$\equiv (\neg p \lor \neg q) \lor p \qquad \text{De Morgan}$$
$$\equiv (\neg q \lor \neg p) \lor p \qquad \text{Commutative Law}$$
$$\equiv \neg q \lor (\neg p \lor p) \qquad \text{Associative Law}$$
$$\equiv \neg q \lor T \qquad \text{Negation Law}$$
$$\equiv T \qquad \text{Identity Law}$$

$\square$

Table 1.17 Logical Operator Precedence

Operator
¬
∧
∨
→
↔

Reaching a value of T on the right side means the statement is a tautology as in this case.

Precedence of Logical Operators
Logical operators have priority as shown below in Table 1.17 with the higher priority operator closer to the top.

For example,

$$\neg p \wedge q \vee r \rightarrow \neg r \equiv (((\neg p) \wedge q) \vee r) \rightarrow (\neg r)$$

1.2 Predicate Logic

We have investigated simple and compound statements that are logically true or false up to this point. However, in many cases, the truth value of a statement depends on the subject. For example, in the sentence "He lives in London", we can not evaluate the correctness of this statement as it truth value depends on who "He" is. Substituting "Kevin" for "He" in this sentence (and assuming we know who Kevin is), we can assert the correctness of this statement. The predicate in this statement is "lives in London".

A *predicate* in a statement describes the subject or other objects and its truth value can be obtained by replacing the subject with a variable. More formally, a predicate is a statement involving variables over a specified set called its *domain*. That is, the domain of a predicate variable is the set of all values that may be substituted for that variable. Thus, substituting a value of a predicate variable from its domain provides us the truth value of a predicate. We are commonly provided with the domain and asked to find the truth value of the predicate. The *truth set* of a predicate is the set of all members of the domain that result in a truth value of the predicate. We show a predicate with capital letters as $P(x)$, meaning P is a predicate with the predicate variable x which is also called the *propositional function P* of x. Truth value of a propositional function depends the value of the variable it refers. For example, if $P(x)$ is "$x > 8$", then substituting 7 for x yields a false value whereas $x = 12$ results in a true value of the function. A propositional function may have more than one variable as in the example below.

Example 1.2.1 Let the propositional function $P(x, y, z)$ to be defined as $z^2 = x + y$. Then, $P(2, 7, 3)$ and $P(5, 11, 4)$ both yield true values whereas $P(3, 2, 4)$ yields a false value.

1.2.1 Quantifiers

Substituting a value for a variable in a propositional function is not the only way to obtain a proposition from such a function. *Quantifiers* can also be used for the same purpose and this process is called *quantification*. In other words, quantifiers specify what part of the domain is to be used when evaluating the correctness of a propositional function. Two main quantifiers are the *universal* and the *existential* quantifiers.

The Universal Quantifier
In various propositional functions, we may need to assert that the value of the proposition is true for all values of the variable in the given domain. For example, "For every integer x, the square of x is a positive integer" is one such function. The universal quantification provides a shorthand representation of this condition and is shown by $\forall x P(x)$. This statement is read as "for all x, $P(x)$" or "for every x, $P(x)$" and the symbol $\forall$ is called the *universal quantifier*.

The Existential Quantifier
The existential quantifier provides another way of quantifying a propositional function. It means "there exists", which in fact means there is at least one variable that provides a true value for the propositional function. This quantifier is shown by the $\exists$ symbol, for example, if $P(x) : x^2 = 1$, $\exists x P(x)$ means there exists at least one variable x that satisfies $P(x)$. In fact, we can see $P(-1)$ and $P(1)$ both provide true values. Other words used for the existential quantifier are "for some x" and "for at least one x".

Example 1.2.2 Let $P(x)$: "x is a student of Computer Science" and $Q(x)$: "x takes Discrete Math course". Let the domain be some university, then,

$$\exists(P(x) \wedge Q(x))$$

means there exists one (there is at least one) Computer Science student who takes a Discrete Math course in that university. Note that we have used the *and* operator between two statements and the existential operator refers to the conjunction of the two statements.

The precedence of quantifiers are higher than all of the logical operators. For example,

$$\exists x P(x) \wedge Q(x) \equiv (\exists x P(x)) \wedge Q(x) \neq \exists x(P(x) \wedge Q(x))$$

1.2.2 Propositional Functions with Two Variables

We can have predicates with more than one variable such as x and y as noted. For example, the predicate "x studies at y University" is such a function which can be specified as $P(x, y)$ where x denotes the person and y is the name of the university. The values of these variables are needed to convert the propositional function to a proposition. However, we may use quantifiers for the same purpose as we have noted. For example, let

$$P(x, y) : y = x + 1$$

be a propositional function with the real numbers as the domain. Without specifying the values of x and y, we can convert this function to a proposition using universal and existential quantifiers. We would normally have four of such combinations since there are two variables but the actual number of combinations is eight as the order of the quantifiers is significant and provides a different meaning when the order is changed. All of the possible eight combinations for the propositional function using quantifiers are listed below.

- $\forall x \forall y P(x, y)$
- $\forall y \forall x P(x, y)$
- $\forall x \exists y P(x, y)$
- $\forall y \exists x P(x, y)$

- $\exists x \exists y P(x, y)$
- $\exists y \exists x P(x, y)$
- $\exists x \forall y P(x, y)$
- $\exists y \forall x P(x, y)$

For example, $\forall x \exists y P(x, y)$ means "for every x there exists a y". For the above proposition, this means for any real number $x \in \mathbb{R}$, we can find at least one $y \in \mathbb{R}$ that is one greater than x which is a proposition with a true value in this case. However, when the order of the quantifiers are reversed, we have $\exists y \forall x P(x, y)$ which means there exists a real number $y \in \mathbb{R}$, for all $x \in \mathbb{R}$ that is one less than x which clearly is false.

Example 1.2.3 Let $L(x, y)$ be "x loves y". Then, $\forall x \exists y$ means "everyone loves at least someone" and $\exists x \forall y$ means "there exists at least one person who loves everyone".

Example 1.2.4 Let domains S be a number of students and P be a number of pizzas. Predicate $E(s, p)$ means student s is eating pizza p. We can then have the following possibilities:

- $\exists s \exists p E(s, p)$: There is (exists) a student s eating a pizza p.
- $\exists p \exists s E(s, p)$: There is a pizza p that student s is eating.
- $\forall s \forall p E(s, p)$: All students are eating pizzas.
- $\forall p \forall s E(s, p)$: All pizzas are are being eaten by all students.
- $\forall s \exists p E(s, p)$: For all students, there exists pizzas That is, they are all eating but there may be some extra pizzas.

- $\exists p \forall s\, E(s, p)$: There exists a pizza (a single one) that all students are eating.
- $\forall p \exists s\, E(s, p)$: Every pizza is being eaten by some student. Note that if the number of students is larger than the number of pizzas, this means some students are left without pizza.
- $\exists s \forall p\, E(s, p)$: There exists a student who is eating all pizzas.

1.2.3   Negation

Negation of universal statements and existential statements need to be considered separately as below.

1.2.3.1   Negation of a Universal Statement

The negation of a universal statement of the form

$$\forall x \in D,\, P(x)$$

which means there exists x in the domain D, $P(x)$ is true, negation of which is logically equivalent to the statement:

$$\exists x \in D,\, \neg P(x)$$

which can be translated as "there exists at least one x in domain D such that negation of $P(x)$ is true". Let us illustrate this concept by an example. Let the universal statement be "All students in Discrete Math class received grade B or higher". Here, the domain is the students enrolled in this class and the proposition is that each student in this class received a grade B or higher. The negation of this universal statement would then be "There exists at least one student in Discrete Mathematics class who received a grade less than B". Let $P(x)$ be "receiving a grade less than B", we can state the following for this above example omitting the domain and the negation is as stated.

$$\neg(\forall x,\, \neg P(x)) \equiv \exists x,\, \neg\neg P(x)$$
$$\equiv \exists x,\, P(x)$$

1.2.3.2   Negation of an Existential Statement

The negation of a universal statement of the form

$$\exists x \in D,\, P(x)$$

which means there exists x in the domain D, $P(x)$ is true, negation of which is logically equivalent to the statement:

$$\forall x \in D,\, \neg P(x)$$

which can be translated as "for all x in domain D, negation of $P(x)$ is true". For example, let us assume that the existential statement to be "There exists one student in Chemistry class who wears glasses". Here, the domain is the students enrolled in Chemistry class and the proposition is that there is at least one student who wears glasses. The negation of this existential statement would then be "All of the students in Chemistry class do not war glasses" or more commonly "There is not a student in Chemistry class who wears glasses". Using the above result, we can form the following.

$$\neg(\exists x, P(x)) \equiv \forall x, \neg P(x)$$

1.2.3.3 Negation of Propositional Functions with Two Variables

The negation of propositional functions with two variables can be done by successively implementing the rules of negation in these functions. For example,

$$\neg(\forall x \exists y P(x, y)) \equiv \exists x \neg(\exists y P(x, y))$$
$$\equiv \exists x \forall y \neg P(x, y)$$

and,

$$\neg(\exists x \forall y P(x, y)) \equiv \forall x \neg(\forall y P(x, y))$$
$$\equiv \forall x \exists y \neg P(x, y)$$

As can be seen, the negation of a propositional function with two variables and two quantifiers results in a propositional function in which the existential quantifier is changed to the universal quantifier and the universal quantifier is changed to the existential one and the propositional function is negated.

1.2.4 The Universal Conditional Statement

A universal conditional statement is a universal statement with a condition and is of the form,

$$\forall x, \text{ if } P(x) \text{ then } Q(x)$$

For example, let $P(x)$: "x is an integer" and $Q(x)$: "square of x is a positive integer", then this proposition can be written as below,

$$\forall x \in \mathbb{Z}, (P(x) \to Q(x))$$

Let us consider the statement "Every integer greater than 1 has a square greater than 3". Forming the propositional functions;

$$\forall x, (x \text{ is an integer} > 1) \to (x^2 > 3)$$

with $P(x)$: "x is greater than 1", and $Q(x)$: "square of x is greater than 3". The negation of a condition is $\neg(p \to q) \equiv p \wedge \neg q$ as was shown in Table 1.15. Thus, the negation of a universal conditional statement can be obtained using this property as follows:

$$\neg(\forall x, P(x) \to Q(x)) \equiv \exists x \neg(P(x) \to Q(x))$$
$$\equiv \exists x(P(x) \wedge \neg Q(x))$$

Considering the above example, the negation of the universal conditional statement would be "there exists an integer greater than 1 and its square is less than 3".

Example 1.2.5 Let the domain be all humans. Write the proposition "All humans are mortal" as a conditional by specifying the statements.

Solution: Let $P(x)$ be "x is human" and $Q(x)$ be "x is mortal". Then,

$$\forall x(P(x) \to Q(x))$$

means for all x variables, if x is a human, then x is mortal.

1.2.5   The Existential Conditional Statements

Similarly, an existential conditional statement is a statement that is both existential and conditional as follows,

$$\exists x, \text{ if } P(x) \text{ then } Q(x)$$

Let us consider $P(x)$ to be "x is an even number greater than 2" and $Q(x)$ as "x is the sum of two prime numbers". Then, the existential conditional statement can be stated as "there exists an even number that is greater than 2 which is the sum of two primes". The negation of an existential conditional statement can be obtained by substitution of the conditional statement as above,

$$\neg(\exists x, P(x) \to Q(x)) \equiv \forall x \neg(P(x) \to Q(x))$$
$$\equiv \forall x(P(x) \wedge \neg Q(x))$$

For the above example, the negated statement would be "for all numbers x, x is an even number greater than 2, and x is not the sum of two primes". In other words, no such number x can be found.

1.3 Review Questions

1. What is the main difference between the propositional logic and predicate logic?
2. What are the main connectives to form compound statements?
3. Give an example of exclusive-or statement.
4. When does a conditional statement yield a false value? Give an example.
5. How can a biconditional statement be verified?
6. Give an example of a biconditional statement yielding a false value.
7. What is the relation between a biconditional statement and a tautology?
8. Give an example of a contradiction.
9. What are the two main methods of proving equivalence of two logical statements?
10. What are De Morgan's laws as applied to logic?
11. What are quantifiers and why are they used?
12. What is the inverse of the statement $\exists x\, P(x)$ and the inverse of the statement $\forall x\, P(x)$.

1.4 Chapter Notes

We have reviewed the main principles of logic in this chapter. A proposition or a statement has either a true or a false value. Propositions can be combined using connectives such as *and* and *or* to form compound propositions. A conditional statement is denoted by $p \to q$ and yields a false value only when p is true and q is false. A biconditional statement is shown as $p \leftrightarrow q$ and has a true value only when p and q have equal truth values. The first method we have reviewed to prove the equivalence of two statements is listing all possible values of the statements in a truth table and checking each row of the value of the statements to be equal. For n statements, the truth table will have 2^n rows making this method difficult to use this when $n > 3$.

Laws of logic provide simplifications and result in an alternative and efficient way of proving that two statements are equivalent. Given two statements, we provide iteratively simpler statements until a known result such as $p \vee T = 1$ is encountered. A predicate provides a description of some entity and its value can be determined with the current entity or object under consideration. A quantifier is used to determine the truth value of a predicate; the universal quantifier declares all instances of the object to be specified and an existential quantifier specifies at least one instance of the object under consideration. We have seen how to negate statements with quantifiers and evaluate conditional statements involving quantifiers.

Exercises

Assume p, q, r are propositions.

1. Let the statement p be "3 is a positive integer" and q be "6 is an odd integer".

    a. Form the conditional $p \rightarrow q$ in words.
    b. Form the conditional $q \rightarrow p$ in words.
    c. Write the biconditional $p \leftrightarrow q$.
    d. Determine the truth value of the statements in (a), (b) and (c).

2. Form the converse, inverse and and contrapositive of each of the following implications.

    a. If it is sunny, I will go out for a walk.
    b. If a is an odd integer, then $a + 1$ is an even integer.
    c. If i is an integer, then i^2 is greater than or equal to 0.

3. Determine whether the following compound propositions are equal using truth tables.

    a. $(p \wedge \neg q) \vee (p \wedge q) = p$.
    b. $p \wedge (q \vee r) = (p \wedge q) \vee (p \wedge r)$.
    c. $p \oplus q = \neg p \oplus \neg q$.

    d. $(p \rightarrow q) = \neg p \vee q$.
    e. $p \vee (q \wedge r) = (p \vee q) \wedge (p \vee r)$.
    f. $\neg (p \oplus q) = \neg p \oplus q$.

4. Use truth tables to verify the following.

    a. $(p \rightarrow q) \vee (p \rightarrow r) \equiv p \rightarrow (q \vee r)$.
    b. $(p \rightarrow r) \vee (q \rightarrow r) \equiv (p \wedge q) \rightarrow r$.
    c. $(p \rightarrow r) \wedge (q \rightarrow r) \equiv (p \vee q) \rightarrow r$.
    d. $(p \vee q) \equiv \neg p \rightarrow q$.

5. Use truth tables to verify the following.

    a. $(p \leftrightarrow q) \equiv \neg p \leftrightarrow q$.
    b. $\neg (p \leftrightarrow q) \equiv p \leftrightarrow \neg q$.

    c. $(p \rightarrow r) \wedge (q \rightarrow r) \equiv (p \vee q) \rightarrow r$.
    d. $(p \vee q) \equiv \neg p \rightarrow q$.

6. Determine which of the following statements are tautologies, contradictions or neither of them.

    a. $p \rightarrow (p \vee q)$.
    b. $(p \rightarrow q) \rightarrow (p \wedge q)$.
    c. $(p \wedge q) \rightarrow p$.
    d. $\neg q \vee (p \rightarrow q)$.

    e. $(p \wedge q) \wedge \neg (p \vee q)$.
    f. $(p \wedge \neg q) \wedge (\neg p \vee q)$.
    g. $((p \leftrightarrow q) \wedge q) \rightarrow p$.
    h. $(p \wedge q) \wedge (q \rightarrow \neg p)$.

7. Prove the following using laws of logic.

a. $p \wedge (p \vee q) \equiv p$.

b. $(p \wedge q) \vee p \equiv p$.

c. $\neg(\neg p \wedge q) \wedge (p \vee q) \equiv p$.

d. $(p \wedge q) \equiv \neg(p \rightarrow \neg q)$.

8. Let $A = \{2, 3, 4, 5, 6\}$. Assess the truth values of the following statements.

a. $\exists x \in A, x^2 = 25$.

b. $\forall x \in A, x + 5 > 6$.

c. $\forall x \in A, x + 4 < 10$.

d. $\exists x \in A, x + 2 \geq 8$.

9. Find the negation of the following statements with the domain of $\mathbb{N}$, determine the truth value of the negation and state the negation in words.

a. $\exists x \forall y, x = y^2$.

b. $\forall x \forall y, x + 4 < 10$.

c. $\exists x \exists y \forall z, z^2 = x^2 + y^2$.

d. $\forall x \exists y, x = -y$.

Proofs

A mathematical system consists of axioms, definitions, theorems and various other structures. A *theorem* is a proposition that can be proved to be true and an argument that establishes the truth of a statement is called a *proof*. Proving a theorem can be accomplished by a direct method or indirectly. Proving propositions that involve quantifiers requires careful reasoning. In this chapter, we review main methods of proof which include direct and indirect methods, proving propositions with quantifiers, proof by cases and review general principles of proofs.

2.1 Arguments

An *argument* consists of a set of propositions p_1, p_2, ..., p_n called *premises* or *hypotheses* followed by a statement q called the *conclusion*. An argument is *valid* if whenever p_1, p_2, ..., p_n are all true, then q is also true. *Inference rules* are simple argument forms that are used to construct more complex arguments.

2.1.1 Rules of Inference

The first rule of inference called *Modus Ponens* is depicted below. It basically means that if $p \rightarrow q$ is valid and if p is true, then we can deduce q is also true. For example, let p be "I study hard" and q be "I will get an A in math". The argument is "if I study hard, I will get an A in Math". "I studied hard" (p is true) implies the second line of this rule which means I will get an A in Math. The main point here is a true statement implying a false statement will yield a false value for the conditional statement, thus, $p \rightarrow q$ having a true value and p having a true value leaves us with the only possibility of q being true.

© Springer Nature Switzerland AG 2021

K. Erciyes, *Discrete Mathematics and Graph Theory*, Undergraduate Topics
in Computer Science, https://doi.org/10.1007/978-3-030-61115-6_2

$$
\begin{array}{ll}
\textbf{Modus Ponens:} & p \\
\text{(mode that affirms)} & p \rightarrow q \\
\hline
 & \therefore q
\end{array}
$$

Modus Tollens states that the inverse of the conclusion implies the inverse of the premise as shown below. We know this rule since $p \rightarrow q \equiv \neg q \rightarrow \neg p$. This rule is the basis of a main proof method called *contrapositive* as we will see. Let p be "It rains" and q be "I get wet". The first statement then is "If it rains, I get wet". If the conditional is true, the Modus Tollens rule states that the fact that I am not wet implies it does not rain.

$$
\begin{array}{ll}
\textbf{Modus Tollens:} & p \rightarrow q \\
\text{(mode that denies)} & \neg q \\
\hline
 & \therefore \neg p
\end{array}
$$

Disjunctive Syllogism states that if at least one of the premises is true and we know one is false, the other one must be true. Let p be "I will study math" and q be "I will go to the movies". The first statement is "I will study math or I will go to the movies". If this statement is true, and the second one is "I will not study math" is valid, then it must be true that I will go to the movies.

$$
\begin{array}{ll}
\textbf{Disjunctive Syllogism:} & p \vee q \\
 & \neg q \\
\hline
 & \therefore p
\end{array}
$$

The *transition rule* (or *hypothetical syllogism*) may seem more obvious then the above rules. Let p be "It rains" and q be "I stay at home" and r be "I study math". This rule states that "if it rains, I stay at home" ($p \rightarrow q$), and "if I stay at home, I study math" ($q \rightarrow r$) statements are both true, then "if it rains, then I study math" ($p \rightarrow r$) statement is also true.

$$
\begin{array}{ll}
\textbf{Transition:} & p \rightarrow q \\
 & q \rightarrow r \\
\hline
 & \therefore p \rightarrow r
\end{array}
$$

The *resolution rule* may best be stated by an example. Let p be "I study math" and q be "I go to the movies" and r be "I meet a friend". The first statement is "I study math or I meet a friend". The second statement is "I go to the movies or I do not meet a friend". Since r cannot be true and *false* at the same time, it must be the case that either p or q is true. In this case "I study math or I go to the movies" is true.

$$\textbf{Resolution:} \quad \begin{array}{l} p \vee r \\ q \vee \neg r \\ \hline \therefore p \vee q \end{array}$$

The *addition rule* says if one of two statements is true, then their conjugate is also true as shown below.

$$\textbf{Addition:} \quad \begin{array}{l} p \\ \hline \therefore p \vee q \end{array}$$

The *simplification rule* means if both of two statements are true, then any of them must be true as shown below.

$$\textbf{Simplification:} \quad \begin{array}{l} p \wedge q \\ \hline \therefore p \text{ and } q \end{array}$$

Finally, *conjunction rule* states that if either of two statements is true, then their conjunction must be true as shown below.

$$\textbf{Conjunction:} \quad \begin{array}{l} p \\ q \\ \hline \therefore p \wedge q \end{array}$$

We can build a *valid argument* by forming a list of statements. Each statement can be a premise or deduced from the preceding statements using rules of inference and the last statement is the conclusion we want to reach. Let us see how we can make use of these rules in a daily example. Let the statements be "I take the train or the bus", and "If I take the train I walk and get tired" and "I am not tired, therefore I took the bus". We need to check whether the last statement is true. Let the statements p be "I take the train", q be "I take the bus", r be "I am tired" The proposition is "I am not tired therefore I took the bus". The premises are $p \vee q$, $q \rightarrow r$ and $\neg r$. The following can be established,

□

Step	Reason
1. $p \vee q$	premise
2. $\neg r$	premise
3. $p \rightarrow r$	premise
4. $\neg p$	2, 3 and modus tollens
5. q	1, 4 and disjunctive syllogism

Example 2.1.1 Given $p \wedge (p \rightarrow q)$, deduce q.
Solution:

Step	Reason
1. $p \wedge (p \rightarrow q)$	premise
2. p	1 and conjunction
3. $p \rightarrow q$	1 and conjunction
4. q	2, 3 and modus ponens

□

Example 2.1.2 Given $(p \vee q)$, $(\neg p \vee r)$ and $(\neg r \vee s)$ deduce $(q \vee s)$.
Solution:

Step	Reason
1. $p \vee q$	premise
2. $\neg p \vee r$	premise
3. $\neg r \vee s$	premise
4. $q \vee r$	1, 2 and resolution
5. $q \vee s$	3, 4 and resolution

□

Example 2.1.3 Given $\neg p \wedge q, r \rightarrow p, \neg r \rightarrow t$ and $t \rightarrow u$, deduce u.
Solution: We start by writing the propositions with their translations.

Step		Reason
1. $\neg p \vee q$ $(\neg p, q)$		premise
2. $r \rightarrow p$ $(\neg r \vee p)$		premise
3. $r \rightarrow t$ $(r \vee t)$		premise
4. $t \rightarrow u$ $(\neg t \vee u)$		premise
5. $\neg r$		resolution of 1 and 2
6. t		resolution of 3 and 5
7. u		resolution of 4 and 6

Applying the resolution law for lines 1 and 2, $q \vee r$ is true. Then, using resolution again for this result with line 3, we have $q \vee s$, the required result.   □

Step		Reason
1. $p \rightarrow q$	premise
2. $r \rightarrow s$	premise
3. $\neg q$	premise
4. r	Premise
5. $\neg p \vee q$	1 translated
6. $\neg r \vee s$	2 translated
7. $\neg p$	3, 5 and resolution
8. s	2 and 4 and modus ponens
9. $\neg p \wedge s$	5, 6 and conjunction

Example 2.1.4 Given $(p \rightarrow q)$, $(r \rightarrow s)$, $\neg q$ and r, deduce $(\neg p \wedge s)$.
Solution: The following steps result in the required proposition.

Step		Reason
1. $p \rightarrow q$	premise
2. $r \rightarrow s$	premise
3. $\neg q$	premise
4. r	premise
5. $\neg p$	1, 3 and modus tollens
6. s	2 and 4 and modus ponens
7. $\neg p \wedge s$	5, 6 and conjunction

□

Example 2.1.5 We could have arrived at the same conclusion using conditional equivalence and resolution principle as shown in above table with more steps to state. □

2.1.2 Definitions

We need to define some terms related to proofs as below.

- *Axiom*: It is a proposition that is assumed to be true.
- *Theorem*: A statement that can be shown to be true using axioms, definitions, other theorems and rules of inference.
- *Lemma*: An intermediate theorem, sometimes called a *helping theorem*, that helps to prove a theorem.
- *Corollary*: A statement that follows an already proved theorem.
- *Conjecture*: A proposition based on incomplete information. A proved conjecture becomes a theorem and a conjecture may be disproved, meaning it may be proved to be false. A disproved conjecture is of not much interest to researchers in general.

We are mostly interested in proving theorems which have the form of a universal statement,

$$\forall x \in D, \quad P(x) \rightarrow Q(x)$$

where D is the domain of the variable x. In general, D is the set of natural numbers, integers, rational numbers or irrational numbers. A starting point to prove a theorem is commonly done by selecting an arbitrary element $u \in D$ and show that $P(u) \rightarrow Q(u)$. This is a sufficient to show $P(x) \rightarrow Q(x)$ for all x as long as u is selected arbitrarily.

A *trivial proof* is the one in which we know q is true regardless of the value of p. A *vacuous proof* is with the statement p being false, $p \rightarrow q$ is true. For example, let p be "It is raining and sunny" and q be "$1 + 1 = 3$", then $p \rightarrow q$ is true.

Example 2.1.6 Let us prove the statement $\forall x \in \mathbb{R}$, if $x > -2$ then $x^2 + 1 > 0$. Since $\forall x \in \mathbb{R}, x^2 \geq 0$; then $\forall x \in \mathbb{R}, x^2 + 1 \geq 0 + 1 > 0$. Thus $Q(x)$ is true $\forall x \in \mathbb{R}$, that is, $x^2 + 1 > 0$ is true for every x value in $\mathbb{R}$, regardless of whether $x > -2$ or not.   □

2.2 Direct Proof

Direct proof method is a straightforward approach where the argument is formed using a series of simple statements. Each such statement follows directly from the previous ones, finally resulting in the proof of the argument. Given the conditional statement of the form "if p then q", direct proof of this statement involves arriving at the conclusion directly from the premises using definitions, axioms, rules of inference and other proven theorems. A general approach for direct proof for a proposition in the form of $\forall x \in D$, if $P(x)$ then $Q(x)$ can be stated as noted.

1. Consider an arbitrary element $u \in D$ that satisfies the premises $P(u)$.
2. Show that the conclusion $Q(u)$ is true using definitions, axioms and the rules of logical inference.

Example 2.2.1 Prove that the sum of two odd integers is even.

Proof We will apply the above steps to prove this proposition. Let us take arbitrary elements $x, y \in D$ which is the set of all odd integers. Since both x and y are odd, we can write $x = 2k + 1$ and $y = 2m + 1$ for some integers k and m. Their sum is $s = 2k + 1 + 2m + 1 = 2(k + m + 1)$ which is an even number since it is divisible by 2.   □

Example 2.2.2 Provide a direct proof of the theorem "If n is an even integer, then n^2 is even".

Proof We need to show $\forall n: n$ is even $\rightarrow n^2$ is even. By the definition of an even integer, $n = 2k$ where k is some integer. Taking square of both sides of this equation yields $n^2 = 4k^2 = 2(2k^2)$ which shows n^2 is twice of some other integer $2k^2$ since $2k^2$ is integer and hence we can conclude n^2 is an even integer.   □

Example 2.2.3 Prove that every odd integer is equal to the difference between the squares of two integers.

Proof An odd integer x can be written as $2k + 1$ for some integer k. Let us check whether this proposition holds for small odd integers.

$$1 = 1 - 0 = 1^2 - 0^2$$
$$3 = 4 - 1 = 2^2 - 1^2$$
$$5 = 9 - 4 = 3^2 - 2^2$$

We observe two things in this sequence; an odd integer is equal to difference between the squares of two consecutive integers, and the sum of these consecutive integers is equal to the odd integer. Now, re-writing x with this guess, let us try to see what happens.

$$x = (k + 1)^2 - k^2$$
$$= (k^2 + 2k + 1) - k^2$$
$$= 2k + 1$$

let $m = (k + 1)^2$ and $n = k^2$, then $x = m^2 - n^2$. □

Note that we can find the values of the two consecutive integers m and n once we write x as $2k + 1$ since $m = k + 1$ and $n = k$. For example, let $x = 41$ which is $2 \times 20 + 1$, thus $m = 21$ and $n = 20$. Checking proves this property since $41 = 21^2 - 20^2 = 441 - 400$. In the other direction, let the two consecutive integers be $m = 53$ and $n = 52$. Since $k = 52$, we can write, $2 \times 52 + 1 = 105 = 53^2 - 52^2 = 2809 - 2704$ which is valid. This example illustrated that sometimes we evaluate the proposition for few input values and try to guess a solution.

Example 2.2.4 Given two integers a and b, if $a + b$ is even then $a - b$ is even.

Proof Since $a + b$ is even, we can write $a + b = 2m$ for some integer m. Then, substitution for b yields:

$$a + b = 2m$$
$$b = 2m - a$$
$$a - b = a - 2m + a$$
$$= 2a - 2m$$
$$= 2(a - m)$$

which shows that the difference is an even number and completes the proof. □

We can have a false proof attempt as in the example below.

$$-2 = -2$$
$$\rightarrow 4 - 6 = 1 - 3$$
$$\rightarrow 4 - 6 + \frac{9}{4} = 1 - 3 + \frac{9}{4}$$
$$\rightarrow \left(2 - \frac{3}{2}\right)^2 = \left(1 - \frac{3}{2}\right)^2$$
$$\rightarrow 2 - \frac{3}{2} = 1 - \frac{3}{2}$$
$$\rightarrow 2 = 1$$

The problem with this false proof is that the square of $-x$ and x are both x^2, thus, the equation in line 4 is correct but taking the square roots to yield the next equation is wrong as the left side is a positive number but the right side is negative. Although direct proof method can be applied with ease in proving various theorems, it is difficult to use this method in many other problems.

2.3   Contrapositive

Proof by contraposition is based on the fact that a conditional statement $p \rightarrow q$ is equal to its contrapositive $\neg q \rightarrow \neg p$ which is basically *modus tollens*. In this case, $\neg q$ is the premise and we attempt to prove $\neg p$ as the conclusion using definitions, axioms, rules of inference and other proven theorems. This method can be applied to a wide range of problems.

Example 2.3.1 Let n be a positive integer and let m divide n. Prove that if n is odd then m is odd.

Proof Proving this theorem directly involves showing m is odd when n is odd and this does not seem obvious. We will try proof by contraposition in which case we need to show that if m is not odd then n is not odd. Our initial condition that m divides n is still valid. If m is not odd, then it is even and hence $m = 2k$ for some integer k. Since m divides n, $n = mr$ for some integer r and substitution for m in this equation yields $n = (2k)r = 2(kr)$ which shows n is even as it is twice of some integer kr. We assumed multiplication of two integers yields another integer. Thus, we proved the contrapositive of the theorem which concludes the proof.   □

Example 2.3.2 Prove that for any integer $a > 2$, if a is a prime number, then a is an odd number.

Proof Let us assume the opposite of the conclusion that a is even. We can then write $a = 2n$ for some integer n. However, this implies a is divisible by 2 and hence it cannot be a prime number which contradicts the premise. We have arrived at the inverse of the premise by assuming the inverse of the conclusion. □

Example 2.3.3 Prove that if the average of three different integers is 8, then at least one of the integers is greater than 9.

Proof Let us form the component propositions p and q. p: "The average of three different integers is 8", q: "One of these integers is greater than 8". Instead of proving $p \to q$, we will attempt to prove the contrapositive $\neg q \to \neg p$. Let us denote the integers by a, b, and c. The inverse of q can be stated as "all of these integers are less than or equal to 8". Let us assume the worst case with the least possible sum where $a + b + c = 6 + 7 + 8 = 21$ in which case their average is 7 which contradicts the premise which states the average is 8 and this completes the proof. □

Example 2.3.4 Prove that if n is an integer and $5n + 4$ is odd, then n is odd.

Proof Let p be "$5n + 4$ is odd" and q be "n is odd". Then $\neg q$ is "n is even" which means $n = 2k$ for some integer k. Substituting in p for n yields, $5n + 4 = 5(2k) + 4 = 10k + 4 = 2(5k + 2)$. Therefore, $5n + 4$ is even as it can be divided by 2 and hence we have proven $\neg p$ is true. □

2.4 Proof by Contradiction

In this proof method, we assume the premise p is true and the conclusion q is not true $(p \wedge \neg q)$ and try to find a contradiction. This contradiction can be against what we assume as hypothesis or simply be something against what we know to be true such as 1=0. In this case, if we find $(p \wedge \neg q)$ is false, it means $\neg(p \wedge \neg q)$ is true. This means either $\neg p$ is true or q is true by De Morgan. We assumed p is true, thus, $\neg p$ is false which ensures q is true. Thus, encountering a contradiction through this process is sufficient to show that q is true and that completes the proof. This method was often practiced by ancient Greek philosophers to solve many interesting problems.

Example 2.4.1 Prove that the sum of two even numbers is even.

Proof Let the numbers be a and b, their sum be c, and the propositions p be "a and b are even" and q be "the sum of a and b, c is even". The inverse of q means c is odd which can then be written as $2k + 1$ for an integer k. Since a is even it can be written as $2n$ for some integer n and similarly $b = 2m$ for some integer m. Then,

$c = 2k + 1 = a + b = 2(n + m)$ which shows that an even number equals an odd number, therefore a contradiction.   □

A *rational* number is a number r that can be expressed as $r = \frac{p}{q}$ where p and q are integers, $q \neq 0$, and p and q have no common divisors other than ± 1. A number that is not rational is called an *irrational* number.

Example 2.4.2 Prove that $\sqrt{2}$ is an irrational number.

Proof This is a classic contradiction proof example. We will assume this statement is *false* and try to arrive at a contradiction. Let us assume $\sqrt{2}$ is a rational number and hence can be written as $\sqrt{2} = a/b$ in simplest form for some integers a and b. These integers can not both be even, otherwise we could have divided them by 2 and arrived at a simpler fraction. Squaring both sides yields $a^2 = 2b^2$, therefore a^2 is even which means a must be even as the square of an odd integer is odd. This means b is odd since a and b can not both be even integers. We can therefore write $a = 2k$ and $b = 2m + 1$ for some integers k and m. Substitution of these values in $a^2 = 2b^2$ yields:

$$4k^2 = 2(2m + 1)^2 = 2(4m^2 + 4m + 1) = 8m^2 + 8m + 2$$
$$2k^2 = 4m^2 + 4m + 1 = 2(m^2 + 2m) + 1$$

which equates an even number on the left to an odd number on the right resulting in a contradiction.   □

2.5   Proving Biconditional Propositions

In order to prove a biconditional statement $p \leftrightarrow q$, we need to show $p \rightarrow q$ and $q \rightarrow p$ since $p \leftrightarrow q = (p \rightarrow q) \wedge (q \rightarrow p)$ and thus validity in both directions is needed. Therefore, proof of a biconditional statement involves solving these two distinct steps. Commonly, the phrase "if and only if" is used between the two statements that make up the biconditional to indicate that the compound statement is a biconditional.

Example 2.5.1 Prove that for any integer n, n^2 is odd if and only if n is odd.

Proof We need to prove in fact two statements:

1. If n is odd, then n^2 is odd
2. If n^2 is odd, then n is odd

Letting p: "n is odd", and q: "n^2 is odd", we will attempt to prove, $p \rightarrow q$ and $q \rightarrow p$. If n is odd, it can be written as $2k + 1$ for some integer k. Then,

$$n^2 = (2k + 1)^2 = 4k^2 + 4k + 1 = 2(k^2 + 2k) + 1$$

Let $m = k^2 + 2k$, then

$$n^2 = 2m + 1$$

and by the definition of an odd number, n^2 is and odd number. In order to prove $q \to p$, we will use the method of contrapositive proof. Let n be even; then, $n = 2k$ for some integer k.

$$n^2 = 4k^2 = 2(2k^2)$$

which means n^2 is even, thus a contradiction with the premise. We have proven the biconditional statement by separately proving $p \to q$ and $q \to p$, using direct proof method for the first, and the contrapositive method for the second. □

Example 2.5.2 Prove that given two integers a and b, their product ab is even if and only if a is even or b is even.

Proof Let p: "a is even or b is even" and q: "product of a and b is even". Hence, we need to prove $p \leftrightarrow q$ in the usual sense. Note that the statement of the problem is of the form $q \leftrightarrow p$ for easiness in understanding. Nevertheless, we need to prove the conditional statements in both directions. Let us first prove $p \to q$ which states that when a is even or b is even, then ab is even. If a is even $a = 2m$ for some integer m, thus, $ab = 2mb$ and is even. Using similar logic for b being even, ab is again even. In the other direction, we need to prove that if ab is even, then a or b is even. Let us use the contrapositive method and assume "a is odd *and* b is odd". Note that this statement is the negation of the statement "a is even or b is even". Then, $a = 2k + 1$ and $b = 2m + 1$ for some integers k and m. Then,

$$ab = (2k + 1)(2m + 1)$$
$$= 4km + 2k + 2m + 1$$
$$= 2(2km + k + m) + 1$$

which shows that the product ab is an odd number. We have shown that $q \to p \equiv \neg p \to \neg q$ is true which completes the proof. □

2.6   Proofs Using Quantifiers

A statement using one quantifier type may be expressed in terms of the other one, that is,

$$\neg(\forall x \in D, P(x)) \equiv \exists x \in D, \neg P(x) \tag{2.1}$$

which is where the counter example proof method comes from and,

$$\neg(\exists x \in D, P(x)) \equiv \forall x \in D, \neg P(x) \tag{2.2}$$

2.6.1 Proving Universal Statements

A universal statement was stated as $\forall x \in D, Q(x)$ is true. Vast majority of mathematical statements are universal statements. Proving such a statement can be done by using the direct proof or other methods described. Alternatively, negation of a universal statement yields an existential statement and disproving the negation of this statement is equivalent to proving it. Essentially, we have three main approaches to prove universal statements:

- *Exhaustion Method*: When D is finite and small, we can check whether $Q(x)$ is true for all $P(x)$. This method clearly is not suitable when size of D is large.

Example 2.6.1 Prove that if n is even and $8 \leq n \leq 16$, then n can be written as the sum of two prime numbers.

Proof We check every even integer in the range, $8 = 3 + 5$; $10 = 3 + 7$; $12 = 5 + 7$; $14 = 7 + 7$; and $16 = 5 + 11$ and hence conclude that the statement is true. $\square$

- *Direct Proof*: This method involves showing $\forall x \in D$, if $P(x)$ then $Q(x)$ as stated.
- *Negation Method*: In this case, we make use of Eq. 2.1 and attempt to find a contradiction. We select a likely u that proves $u \in D$, prove $\neg P(u)$ which means $\forall x \in D, P(x)$ is true. This method is also called *proof by counter examples*.

Example 2.6.2 Prove that $\forall x \in \mathbb{R}, x^2 + 1 > 0$.

Proof Negation of this statement yields: $\exists x \in \mathbb{R}, x^2 + 1 \leq 0$. But we know by the definition that the square of any number is positive hence it is not the case that $x^2 \leq -1$ meaning a contradiction. $\square$

Example 2.6.3 For all positive prime numbers, if p is prime then $2p + 1$ is also prime.

Proof Let $p = 7$ which is a prime number. But, $2p + 1 = 15$ is not prime. $\square$

Example 2.6.4 All numbers between 4 and 12 are prime.

Proof Let $x = 9$, it is between 4 and 12 but not prime. $\square$

Two rules may be stated for universal statements as in existential statements:

- *Universal Instantiation*: If $\forall x \in D$, $P(x)$ is stated, then $P(u)$ is true when $u \in D$. Let the statement be "All countries have capitals", and since Portugal is an instance of the domain countries, it must have a capital.
- *Universal Generalization* : If $P(u)$ is true for an arbitrarily chosen $u \in D$ then we can generalize that $\forall x \in D$, $P(x)$. The element u must not be specific to arrive at the universal generalization. For example, we select an *arbitrary* monkey and observe it has two legs. We can then generalize and say "all monkeys have two legs".

2.6.2 Proving Existential Statements

A statement of the form $\exists x \in D$, $P(x)$ was named an existential statement in Chap. 1. This statement is true if and only if $P(x)$ is true for at least one explicit $x \in D$. A simple way to prove such a statement is to find at least one x that makes $P(x)$ true. Alternatively, a set of directions that yield such x values can be stated. Both approaches are called *constructive proofs of existence*. We can have direct or negation method to prove existential statements.

2.6.2.1 Direct Method
We attempt to deduce $Q(x)$ from $P(x)$ by finding an element that satisfies P as shown in the examples below.

Example 2.6.5 Prove that there exists an even integer n that can be written as the sum of two prime numbers.

Proof The universal form of this statement is the *Goldbach Conjecture* which is not proven to date. Let $n = 18$, which is the sum of 5 and 13 which are both prime numbers. $\square$

Example 2.6.6 Prove that $\exists x \in \mathbb{R}$, $x^2 < x$.

Proof If we take $x = \frac{1}{2}$ then $x^2 = \frac{1}{4} < \frac{1}{2} = x$. $\square$

Example 2.6.7 Prove that $\forall x \exists y \in \mathbb{N}$, $y = x^2$.

Proof For any $x \in \mathbb{N}$, its square is also in $\mathbb{N}$. $\square$

2.6.2.2 Negation Method
The *negation method* to prove an existential statement can be performed by negating the statement and then searching for a contradiction as in the example below. We start with $\forall x \in D$, $\neg P(x)$ and proceed to derive a contradiction.

Example 2.6.8 Prove $\exists x \in \mathbb{R} : 0 \leq x < 1$.

Proof Negation of this statement is $\forall x \in \mathbb{R} : (x < 0) \vee (x \geq 1)$. This statement is clearly false, take $x = 0.4$ for example.   □

Two rules may be stated for existential statements:

- *Existential Instantiation*: If $\exists x \in D$, $P(x)$ is stated, then $P(u)$ is true for some $u \in D$. Let the statement be "There exists carnivore plants", therefore there is a plant x that is carnivore.
- *Existential Generalization* : If $P(u) \in D$ is true for some element u, then $\exists x \in D$ $P(x)$. Let the first statement be "Miguel in our class can speak Spanish", then we can say "There exists someone in our class who can speak Spanish".

We can generalize the rules of inference for quantified statement as shown below.

Universal Instantiation:	$\dfrac{\forall x P(x)}{\therefore P(u)}$
Universal Generalization:	$\dfrac{P(u) \text{ for an arbitrary } u}{\therefore \forall x P(x)}$
Existential Instantiation:	$\dfrac{\exists x P(x)}{\therefore P(u) \text{ for some element } u}$
Existential Generalization:	$\dfrac{P(u) \text{ for some element } u}{\therefore \exists x P(x)}$

2.7 Proof by Cases

Proof by cases is typically implemented when the premise can be specified as a number of cases. The premise can then be written as a conditional statement of the form,

$$((p_1 \vee p_2 \vee ... \vee p_n) \rightarrow q) \tag{2.3}$$

which is equivalent to

$$((p_1 \rightarrow q) \wedge (p_2 \rightarrow q)... \wedge (p_n \rightarrow q)) \tag{2.4}$$

with each implication being a case. Let us assume q is false and let all p_i be false. The statements in Eqs. 2.3 and 2.4 are both true and thus are equal. If there exists some p_i that is true, then $(p_1 \vee p_2 \vee ... \vee p_n)$ and thus Eq. 2.3 has a false value as

Eq. 2.4. When q is true, then both of these equations give true values and are equal. Therefore, proving each case is equivalent to solving the first statement.

Example 2.7.1 Prove that $\forall n \in \mathbb{Z}, n^2 \geq n$.

Proof We have three cases, $n < 0$, $n = 0$ and $n > 0$.

- Let $n = 0$, then $0 \geq 0$ is true.
- Let $n < 0$, in this case, n^2 is a positive number, thus, $n^2 \geq n$.
- The case $n > 0$ means $n \geq 1$ and multiplying each side by n yields $n^2 \geq n$; since n is positive, the direction of the inequality does not change.

□

Example 2.7.2 Let $n \in \mathbb{Z}$, then $n^2 + n$ is even.

Proof We have two cases, n is even or odd.

- Let n be even. In this case, $n = 2k$ for some integer k. It follows,

$$n^2 + n = (2k)^2 + 2k = 4k^2 + 2k$$
$$= 2(2k^2 + k)$$

- Let n be odd. In this case, $n = 2k + 1$ for some integer k. It follows,

$$n^2 + n = (2k + 1)^2 + 2k + 1 = 4k^2 + 4k + 1 + 2k + 1$$
$$= 4k^2 + 6k + 2 = 2(2k^2 + 3k + 1)$$

since n can be expressed as $2m$ for some integer m in both cases, $n^2 + n$ is even. □

2.8 Review Questions

1. What is an axiom?
2. What is the difference between a theorem, a lemma and a corollary?
3. Give an example of modus ponens from everyday life.
4. Give an example of modus tollens from everyday life.
5. Give an example of disjunctive syllogism from everyday life.
6. What is a trivial proof? Give an example.
7. What is a vacuous proof? Give an example.
8. Compare the contrapositive and contradiction methods of proof. How do they differ?

9. What is the difference between proving a conditional statement and proving a biconditional statement?
10. Describe negation method of proof as applied to proving universal and existential statements.
11. When is proof by cases method is used?

2.9 Chapter Notes

Having reviewed the main methods of proof which are direct, contrapositive and contradiction, we may need to know what to do when confronted with proving a statement. A general rule of thumb is to check whether a direct proof is possible. If this is not possible, contrapositive and contradiction methods may be sought.

Proving biconditional statements such as $p \leftrightarrow q$ requires proving both $p \rightarrow q$ and $q \rightarrow p$ as two distinct steps. The basic direct, contrapositive and contradiction methods may be used in these steps as noted. Proofs using quantifiers require careful consideration. We can make use of negation rules of quantifiers; from universal quantifier to existential quantifier and vice versa. In Chap. 6 we will see a powerful method of proof called induction which is frequently used to prove correctness of algorithms.

Exercises

1. Let the statements p be "all of the family is going to the movies", and q be "Jasmine is a member of the family". Show that Jasmine is going to the movies using validity argument method and rules of inference.
2. Prove the following using the rules of inference.

    a. $p \rightarrow q$, $p \wedge r$; then q.   c. $p \rightarrow q$, $r \rightarrow s$, $\neg q, r$; then $\neg p \wedge s$.
    b. $p \rightarrow \neg q$, $q \vee r$; then $p \rightarrow r$.   d. $p \wedge (q \vee r)$, $\neg p \vee \neg q$; then $r \vee q$.

3. Prove that if n is an integer, n^3 is even if and only if n is even.
4. Show that the product of an even integer with any integer is even.
5. Prove that for any two integers a and b, $\min(a, b) + \max(a, b) = a + b$.
6. Show that when one of the two integers is an even and the other is an odd number, their sum is and odd number.
7. Prove that the sum of two consecutive integers is an odd integer.
8. Prove that for all integers a and b, if $a^2 + b^2$ is odd, then a or b is odd.
9. Prove that if n is odd, $2n + 3$ is also odd using the direct proof method.
10. Prove that if n is even, $7n + 4$ is also even using the contrapositive method.
11. Prove that if n is odd, $3n + 5$ is even using the contrapositive method.
12. Use contradiction method to prove that for any positive integer n, if a^n is even, then a is even.
13. Given two integers a and b, prove that ab is odd if and only if a and b are both odd.

14. Use contrapositive method to prove the following: Given to positive integers, if their product is greater than 100, then at least one of the numbers is greater than 10.

15. Prove that if n is an odd integer, $2n + 4$ is an odd integer. Test whether this statement can be modified to be a biconditional statement.

16. Prove that given an integer n, n is even if and only if $n - 1$ is odd.

17. Prove that if n is an even integer, then $5n^3$ is an even integer.

18. Let n be an integer. Then $3n + 8$ is odd if and only if n is odd.

19. Disprove that if n is an even integer then $3n + 6$ is odd.

20. Disprove the statement $\forall x, y \in \mathbb{R}$, if $a^2 < b^2$, then $a < b$.

21. Prove by cases that $|a + b| \le |a| + |b|$.

22. Prove that every prime number greater than 3 is either one more or one less than a multiple of 6.

Algorithms

3

Algorithms have been used for a long time, long before the invent of computers, to solve problems in a structured way. An algorithm is a finite set of instructions or logic, written in order, to accomplish a certain predefined task. There are certain requirements from any algorithm; it may or may not receive inputs but some form of output, which is the solution to the problem at hand, is expected. For example, if we want to find the sum of first n positive integers, n is the input to the algorithm, and the sum is the output. Clearly, the steps of the algorithm should be precise without any ambiguities. An algorithm should terminate after a number of steps and above all, it should provide the correct result. However, finding the correct answer only is not adequate; the output should be provided with minimum number of steps. This would mean an algorithm that works in a shorter time is preferred to another one that gives the same result but works longer.

We start this chapter with basics of algorithms including fundamental algorithm structures and then continuing with basic data structures used in algorithms. The last part of the chapter reviews to main methods of algorithm design.

3.1 Basics

Every Algorithm must satisfy the following properties:

- *Input*: There should be 0 or more inputs supplied externally to the algorithm.
- *Output*: There should be at least 1 output obtained. This output will be the solution to the problem that is investigated.

© Springer Nature Switzerland AG 2021
K. Erciyes, *Discrete Mathematics and Graph Theory*, Undergraduate Topics
in Computer Science, https://doi.org/10.1007/978-3-030-61115-6_3

Fig. 3.1 Block diagram of a
computer

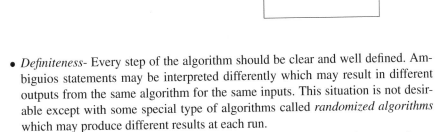

- *Definiteness*- Every step of the algorithm should be clear and well defined. Ambiguios statements may be interpreted differently which may result in different outputs from the same algorithm for the same inputs. This situation is not desirable except with some special type of algorithms called *randomized algorithms* which may produce different results at each run.
- *Finiteness*: The algorithm should have finite number of steps that is, we are interested in algorithms that find the results in a finite interval of time. There are cases however, that the algorithm ends in a finite time but this time interval is so large that it can be considered as infinite for all practical purposes.
- *Correctness*: Every step of the algorithm must generate a correct output. After all, a wrong output is of no use.

An algorithm is said to be efficient and fast, if it takes less time to execute and consumes less memory space than any other algorithm that solves the *same* problem. A *program* has a different meaning than an algorithm, it has the format that can be easily understood and executed by a computer. Moreover, a program may not terminate such as a network program that always waits for input from the network. A computer *programming language* is used to transfer an algorithm to a format that will be understood by the computer.

The block diagram of a computer is depicted in Fig. 3.1 with three essential components: the processor, the memory and the input/output unit. The processor is the brain of the computer where each instruction is executed possibly on some data. Instructions and data reside in memory and the main task of the processor is to fetch instruction and data, execute the instruction and produce some output to be delivered to the output unit. The input unit provides interface between the external world and the processor to transfer input data. A computer may be envisioned as a system that transfers raw data into refined, processed data.

The processor is where the algorithm is executed, the code and data reside in memory in the so-called *Von Neumann Model* of computation. The input/output units are used to communicate with the external world. A computer algorithm commonly works on variables that are stored in memory to provide the desired output.

3.1.1 Pseudocode Convention

An algorithm may be described using various methods. A common and a popular way to specify an algorithm is using *pseudocode* which is a type of structured language. The pseudocode shown in Algorithm 3.1 consists of lines of instructions which are written in plain English and we can see that each step is precisely described. Two integers a and b are input to the algorithm in line 1, their sum is calculated and stored in another integer c in line 2 and finally, the value of c, which is the result of the algorithm, is output. We have 3 precisely defined instructions in this algorithm with the first line as the input, second one as the processing and the last line as the output. This flow is typical in an algorithm as stated. Commonly, we would specify the input parameters to an algorithm and the outputs from it in separate lines at the beginning of the algorithm. Lastly, there is no doubt this algorithm is efficient since we cannot have another shorter algorithm such as 1 or 2 steps to do the same calculation.

Algorithm 3.1 *Sum of integers*

1: **input** two integers a and b
2: let integer $c = a + b$
3: **output** c

3.1.2 Assignment and Types of Variables

A parameter in an algorithm is in the form of a *variable* that can have a value assigned to it as in algebra. Variables have names and a value can be assigned to a variable using "=", ":=" or $\leftarrow$ operators. In many computer programming languages, the standard variable types are the *integer*, *real*, *booelan* and *character* as the names suggest. Many programming languages require explicitly stating of the type of a variable before its use such as "a: integer", or "int a". The following are valid assignments resulting in the storage of value 6 in b when executed consecutively.

$$a \leftarrow 3$$

$$b \leftarrow (a + 9)/2$$

3.1.3 Decision

An algorithm may need to perform a specific action depending on whether a condition is true or even branch to a different address than the usual sequence depending on a condition. The *if* statement is used for this purpose and if the statement after *if* yields a true value, the next statement is executed. Otherwise, the next statement is not executed and the program follows the instruction after the *if* instruction. The *else* control when used causes the statement after else to be executed when if condition results in a false value. Algorithm 3.2 inputs two integers and finds the greater of them and the *print* command is used to provide output. Note that we do not need to test the equal condition since the flow of execution arriving at line 7 of the algorithm means that this is the only possibility.

Algorithm 3.2 *If* statement

1: **Input**: a, b: integer
2: **Output**: the greater of two inputs
3: **if** $a > b$ **then**
4:     **print** a
5: **else if** $b > a$ **then**
6:     **print** b
7: **else**
8:     **print** "equal"
9: **end if**

3.1.4 Loops

A *loop* in an algorithm is used to perform the same operation many times possibly on different data. A loop typically has a control statement and the body of the loop. The control statement is tested to enter the body of the loop or not. Three main types of loops in a computer algorithm are the *while*, *for* and the *repeat ... until* loops described below.

- *while* loop: A *while* loop tests a condition and executes the loop (next statement to the *while* statement) when this condition yields a true value. Otherwise, loop is exited. After execution of the *while* loop, the condition is tested again. An example to add the input numbers until a 0 is entered is shown in Algorithm 3.3. Note that we need two input statements, the first one to have the loop working for the first time with a valid input and the second one to test the condition with the next input.
- *for* loop: A *for* loop is used to iterate the body of a loop. The control statement at the header has a *loop variable* which is initialized as the first part of the statement followed by the test condition. The final part of the control statement specifies the operation on the loop variable at the end of the loop body. The loop in Algorithm 3.4

Algorithm 3.3 *while* loop

1: **Input**: *num*: integer
2: **Output**: *sum*
3: *sum* ← 0
4: **input** *num*
5: **while** *num* ≠ 0 **do**
6:     *sum* ← *sum* + *num*
7:     **input** *num*
8: **end while**

---

has i as the loop variable which is tested at the beginning of each iteration. An example algorithm to calculate the nth power of an integer a is shown in Algorithm 3.4. We observe that this algorithm has inputs a and n, and it multiplies a, n times by itself to find a^n using a *for* loop. We do not specify how the output is presented; it may be returned to a program or may be simply output to the screen.

Algorithm 3.4 *Power of an integer*

1: **Input**: *a*: integer, *n*: integer
2: **Output**: $pow = a^n$
3: *pow* ← 1
4: **for** $i = 1$ to n **do**
5:     *pow* ← *pow* · *a*
6: **end for**

A slightly different form of for loop is the *for all* structure in which we do not specify a control statement with an index and a test condition but state that the loop continues until all elements of a set are processed. The code segment below reads each element a of a group of variables in a set A and outputs them.

for all $a \in A$ **do**

output a

- *repeat .. until* loop: Different than the for and while loops which may not enter the loop based on the test condition, the *repeat .. until* loop is executed at least once. The logical statement after *until* is tested at the end of the loop and a return to repeat line is made if this statement yields a *true* value. Let us rewrite Algorithm 3.3 using the *repeat ... until* loop structure this time as in Algorithm 3.5. Note that we only need to have one input statement this time. Also, test condition is reversed, that is, we repeat the loop until *num* equals 0 and 0 is added to the sum.

Algorithm 3.5 *repeat until* loop

1: **Input**: *num*: integer
2: **Output**: *sum*
3: *sum* ← 0
4: **repeat**
5:     *sum* ← *sum* + *num*
6:     **input** *num*
7: **until** *num* = 0

3.1.5   Functions and Parameter Passing

Some section of an algorithm may be used frequently and writing this part repeatedly results in an algorithm that is long and difficult to analyze. A *procedure* or a *function* is a subprogram that may be invoked many times possibly with different inputs. A procedure is more general and a function is perceived as a procedure that returns a value. A procedure or a function has a list of input parameters and the type of variable returned. Sending its input parameters to a procedure can be performed by the following methods.

- *Call by value*: This is a commonly used method in which the sent values from the main program are copied to the local variables of the procedure.
- *Call by reference*: There are cases when we want the procedure/function to change the value of a parameter passed to it. In such cases, the address of the parameter is passed to the procedure to enable it to modify the value.

Algorithm 3.6 depicts call-by-value invoking of function *Add* which copies the two integers *a* and *b* to its local variables *x* and *y* and returns their sum to the main program. The *Swap* function receives the addresses of integers *a* and *b*, and uses a temporary variable *temp* to swap their contents. The contents of addresses are shown by asterisks and the main program uses the ampersand sign to send the addresses of the variables. The output of this program would be 8, and 5 and 3.

3.2   Basic Data Structures

The standard data types such as integer, real and character may be used to build more complicated data structures. One such structure is an *array* that consists of homogenous data elements. An array has a name, a dimension and its type stated when declared. For example,

$$A[10] : integer$$

Algorithm 3.6 *Sum of integers*

```
1: procedure ADD(x:integer, y:integer)
2:    return(x + y)
3: end procedure
4:
5: procedure SWAP(ap:integer address, bp: integer address)
6:    temp: integer
7:    temp = *ap
8:    *ap = *bp
9:    *bp = temp
10: end procedure
11:
12: main program
13:    a ← 3
14:    b ← 5
15:    print Add(a,b)
16:    Swap(&a, &b)
17:    print a, b
```

is the declaration of array A with 10 integer elements. Accessing an element of an array is done by specifying an index. For example,

$$A[2] \leftarrow 3$$

Algorithm 3.7 places the square of each index in the corresponding array entry. For example, $A[2] = 4$ and $A[9] = 81$ after executing the algorithm.

Algorithm 3.7 *Array example*

```
1: A[10]:integer
2: for i = 1 to n do
3:    A[i] ← i · i
4: end for
```

A two or more dimensional arrays are used to store data in a more structured way. A two dimensional array, commonly called a *matrix*, has elements in its rows and columns. A matrix M with 4 rows and 5 columns of integers may be defines as follows:

$$M[4, 5] : integer$$

Accessing an element of such a matrix requires the row and column of the element to be specified. Algorithm 3.8 shows how the elements of a matrix M are computed by adding the row index i to the column index j to form the contents of the location (i, j) of M. Note that we used two nested for loops, when $i = 1$, j changes from 1 to 5 in the inner loop to store data matrix locations $M[1, 1]$ to $M[1, 5]$. In the next outer loop iteration when $i = 2$, matrix locations $M[2, 1]$ to $M[2, 5]$ are selected. Total number number of execution of line 4 of this algorithm is $4 \times 5 = 20$.

Algorithm 3.8 *Two dimensional array*

1: $M[4, 5]$:integer
2: **for** $i = 1$ to 4 **do**
3:     **for** $j = 1$ to 5 **do**
4:         $M[i, j] \leftarrow i + j$
5:     **end for**
6: **end for**

The formed matrix after the running of this algorithm will have the following elements as below,

$$
\begin{array}{c}
\quad\ 1\ 2\ 3\ 4\ 5 \\
\begin{array}{c} 1 \\ 2 \\ 3 \\ 4 \end{array}
\begin{pmatrix}
1 & 3 & 4 & 5 & 6 \\
2 & 4 & 5 & 6 & 7 \\
3 & 5 & 6 & 7 & 8 \\
4 & 6 & 7 & 8 & 9
\end{pmatrix}
\end{array}
$$

A heterogeneous data structure which consists of different type of data elements may be formed by the *structure* declaration as below,

```
structure student {
            age: integer
            gender: character
            GPA: real } Gabriel;
```

The structure student has integer, character and real fields and declares variable Gabriel of the type of this structure. Accessing a member of such a structure can be performed as below,

$$Gabriel.age = 19$$

is a valid statement that places 19 in the age field of *Gabriel* variable. We can have an array of structures as below,

$$class[20] : \text{structure } student$$

which declares a class of 20 students each with fields of the student structure. Accessing a field of an element can be done by,

$$class[2].GPA = 3.82$$

which places 3.82 as the GPA of the second student of the class.

Example 3.2.1 Write an algorithm in pseudocode that lists the numbers and names of students who have a GPA of 3.0 or higher, using the array *class* in the above example. Also, provide the average of the students who have GPAs of 3.0 or higher. *Solution*: We need to check each entry of the array class using some loop structure. The *for* loop is appropriate as we know how many elements of the array to be searched. The pseudocode of the final algorithm shown in Algorithm 3.9 prints the number and name of students of the required criteria.

Algorithm 3.9 *Structure example*

1: *class*[*n*]:structure student, *i*:integer, *count*: integer, *sum*:real
2: *sum* ← 0.0
3: *count* ← 0
4: **for** *i* = 1 to *n* **do**
5:    **if** *class*[*i*].*GPA* < 3.0 **then**
6:       *count* ← *count* + 1
7:       *sum* ← *sum* + *class*[*i*].*GPA*
8:       **print** *class*[*i*].*num*, *class*[*i*].*name*, *class*[*i*].*GPA*
9:    **end if**
10: **end for**
11: *ave* ← *sum*/*count*
12: **print** *ave*

A structure may have an array embedded in it as,

```
structure book {
            ISBN: integer
            pages: integer
            prices[5]: real } Kafka;
```

which denotes 5 prices in 5 countries for the *book* structure. Assigning a price for the second country can be done as below,

$$Kafka.prices[2] = 32.95$$

3.3 Sorting

Sorting is the process of sequencing a list of elements from higher to lower or vice versa and it is one of the most frequently used algorithms performed by computers. Let us consider an array $A = [3, 2, 0, 8, 4, 1]$ with the aim of sorting elements of this structure from larger to smaller so that $A = [8, 4, 3, 2, 1, 0]$ is obtained at the end of the algorithm. We will look at two basic algorithms for this purpose.

3.3.1 Bubble Sort

The idea of the bubble sort algorithm is to move the largest element to the top of the list at each iteration as the bubbles of boiling water. The first iteration finds the largest element of the array and places it in the nth place, the second iteration finds the second largest and stores it in $(n - 1)$th location and so on. Starting from the first location, each element is compared with its neighbor and if the neighbor has a smaller value, the values are exchanged. The running of the first iteration of this algorithm in a sample sequence is shown below. At the end of the first pass, the largest value 8 is placed at the highest array entry.

$$3 \quad 2 \quad 0 \quad 8 \quad 4 \quad 1$$

$$2 \quad 3 \quad 0 \quad 8 \quad 4 \quad 1$$

$$2 \quad 0 \quad 3 \quad 8 \quad 4 \quad 1$$

$$2 \quad 0 \quad 3 \quad 8 \quad 4 \quad 1$$

$$2 \quad 0 \quad 3 \quad 4 \quad 8 \quad 1$$

$$2 \quad 0 \quad 3 \quad 4 \quad 1 \quad \mathbf{8}$$

The second pass starts from the first element but goes up to $(n - 1)$th element of the array this time since the nth element is determined.

$$2 \quad 0 \quad 3 \quad 4 \quad 1 \quad \mathbf{8}$$

$$0 \quad 2 \quad 3 \quad 4 \quad 1 \quad \mathbf{8}$$

$$0 \quad 2 \quad 3 \quad 4 \quad 1 \quad \mathbf{8}$$

$$0 \quad 2 \quad 3 \quad 4 \quad 1 \quad \mathbf{8}$$

$$0 \quad 2 \quad 3 \quad 1 \quad \mathbf{4} \quad \mathbf{8}$$

From this example we can see that we need two nested for loops with the outer loop running n times and the inner loop running n, $(n - 1)$, $(n - 2)$,...,1 times. A simple way to realize this structure is to have the outer loop with index starting from n and decrementing at each step and the inner loop having the upper limit of the loop as the current value of the outer loop as shown in Algorithm 3.10. The running of the outer loop requires n steps and the inner loop runs at most n steps, thus, the time needed for this algorithms is n^2.

Algorithm 3.10 *Bubble Sort*

1: **Input**: $A[10]$:integer
2: **Output**: sorted A
3: **for** $i = n$ down to 1 **do**
4:     **for** $j = 1$ to (i) **do**
5:         **if** $A[j] < A[j + 1]$ **then**
6:             **swap** $A[j]$ and $A[j + 1]$
7:         **end if**
8:     **end for**
9: **end for**

3.3.2 Exchange Sort

Exchange sort takes a different approach by comparing the first element of the array with all others and swapping the values if the compared value is larger. This way, the first element of the array contains the largest element after first pass. The first iteration of this algorithm in the same sequence is shown below where the largest element 8 is placed in the first location of the array at the end.

$$3 \quad 2 \quad 0 \quad 8 \quad 4 \quad 1$$
$$|--|$$
$$3 \quad 2 \quad 0 \quad 8 \quad 4 \quad 1$$
$$|---|$$
$$3 \quad 2 \quad 0 \quad 8 \quad 4 \quad 1$$
$$|-----|$$
$$8 \quad 2 \quad 0 \quad 3 \quad 4 \quad 1$$
$$|-------|$$
$$8 \quad 2 \quad 0 \quad 3 \quad 4 \quad 1$$
$$|---------|$$
$$\mathbf{8} \quad 2 \quad 0 \quad 3 \quad 4 \quad 1$$

The second pass starts from the second location of the array and compares the value in the second location with all higher index elements by swapping the values if the compared value is larger as shown below.

$$\begin{array}{cccccc} \mathbf{8} & 2 & 0 & 3 & 4 & 1 \\ & |\!-\!-\!| & & & & \\ \mathbf{8} & 2 & 0 & 3 & 4 & 1 \\ & |\!-\!-\!-\!| & & & & \\ \mathbf{8} & 3 & 0 & 2 & 4 & 1 \\ & |\!-\!-\!-\!-\!-\!| & & & & \\ \mathbf{8} & 4 & 0 & 2 & 3 & 1 \\ & |\!-\!-\!-\!-\!-\!-\!-\!| & & & & \\ \mathbf{8} & \mathbf{4} & 0 & 3 & 2 & 1 \end{array}$$

We compare element 1 with 2,...,6; then element 2 with 3,...,6; and then element 3 with 4, 5, 6 and so on. The upper limit is always n but the starting index is incremented at each iteration. This algorithm requires two nested for loops, this time, the inner loop starting index is the current outer loop index and both loops have n as upper index limit as shown in Algorithm 3.11. It requires n^2 steps at most as the bubble sort algorithm.

Algorithm 3.11 *Exchange Sort*

1: **Input**: $A[10]$:integer
2: **Output**: sorted A
3: **for** $i = 1$ to n **do**
4:     **for** $j = i$ to n **do**
5:         **if** $A[i] < A[j]$ **then**
6:             swap $A[j]$ and $A[j + 1]$
7:         **end if**
8:     **end for**
9: **end for**

3.4 Analysis

A fundamental question commonly asked when designing an algorithm is whether a *better* algorithm than the one designed exists. A better algorithm means it works faster, that is, it requires less number of steps. Counting the number of steps that an algorithm requires careful consideration. Let us analyze an algorithm that finds the maximum element of an integer array A of size n depicted in Algorithm 3.12.

This algorithm starts by copying the first element of the array to the variable *max* which will contain the largest element in the end. Then, each element of the array is compared with the value of *max* and any larger value is copied to *max*. Assignment step at line 2 is performed once and can be ignored for any large n value. The *for*

Algorithm 3.12 *Finding max value*

1: **Input**: $A[10]$:integer
2: **Output**: *max*: maximum value of A
3: $max \leftarrow A[1]$
4: **for** $i = 2$ to n **do**
5:    **if** $max < A[i]$ **then**
6:       $max \leftarrow A[i]$
7:    **end if**
8: **end for**

---

loop is executed $n - 1$ times, therefore, we can say this algorithm takes exactly $n - 1$ steps, or n steps when the initial assignment is considered.

Now, let us look at a procedure that searches for a given value a in an array A and returns the index of the element if found as shown in Algorithm 3.13. Array A is searched by comparing each of its element with a until a is found. This time, we do not know how many times *for* loop will be executed. The value we search may be the first element of the array, or somewhere before the end of the array, or it may not exist in the array at all. In the first case, there will be only one step running of the loop and we need n executions in the last case. Any running time between 1 and n inclusive is possible. However, we do know that this algorithm requires at least one step which is its *best execution time* and at most n step called its *worst execution time*.

Algorithm 3.13 *Searching an array*

1: **procedure** SEARCH($A[n]$:integer, a:integer)
2:    **for** $i = 1$ to n **do**
3:       **if** $a = A[i]$ **then**
4:          **return** i
5:       **end if**
6:    **end for**
7:    **return** *not_found*
8: **end procedure**

We are mostly concerned with the worst time complexity of an algorithm as it shows us what to expect in the worst case. Also, we will assume n is large while assessing the number of steps required for the algorithm to finish. For example, let the worst time complexity of an algorithm is computed as $3n^2 + 2n + 5$ steps. First thing to note is that we can discard any constant terms because they have very little effect on running time when n is large. Moreover, the term $2n$ will have a diminishing effect as n increases resulting in a worst time complexity of $3n^2$ or simply n^2, using the same reasoning.

We are now ready to have some formalism to be able to deduce time complexity of an algorithm. Let f and g be two functions from $\mathbb{N}$ to $\mathbb{R}$. The *worst*, *best* and *exact* time complexity of an algorithm may be defined as follows.

- *Worst time complexity (Big-O Notation)*: $f(n) = O(g(n))$, if there exists a constant $c > 0$ such that $f(n) \leq c.g(n)$, $\forall n \geq n_0$ for some n_0. In other words, if we can find a function $g(n)$ that is an upper bound which is always greater than the running time of the algorithm after some threshold value n_0, we can say $g(n)$ is the worst time complexity of the algorithm shown by $O(g(n))$. In our array search example, the worst time complexity is $O(n)$.

Example 3.4.1 Let $f(n) = 4n + 3$. We can then guess $g(n) = n$ since this function approaches n for very large n. Thus,

$$4n + 3 \leq cn$$
$$(c - 4)n \geq 3$$
$$n \geq \frac{3}{c - 4}$$

Selecting $c = 5$ and $n_0 = 3$ satisfies this inequality and we can say that $O(n)$ is the worst time complexity of this function.

- *Best time complexity (Big-Omega Notation)*: $f(n) = \Omega(g(n))$, if there exists a constant $c > 0$ such that $f(n) \geq cg(n)$, $\forall n \geq n_0$ for some n_0. This means if we can find a function $g(n)$ that is always less or equal to the execution time of our algorithm after some threshold value n_0, we can say $g(n)$ is the best time complexity of the algorithm shown by $O(g(n))$. In our array search example, the best time complexity is $O(1)$ since we could find the searched entry in 1 step at best.

Example 3.4.2 Let us consider the algorithm with the running time $f(n) = 4n + 3$ again for best time complexity. If we select $c = 1$ and $n_0 = 1$, then for any $n \geq 1$, $4n + 3 > n$, which means $\Omega(n)$.

- *Big-Theta Notation*: $f(n) = \Theta(g(n))$, if $f(n) = O(g(n))$ and $f(n) = \Omega(g(n))$. That is, if there are two constants $c_1 > 0, c_2 > 0$ with a threshold n_0 and a function $g(n)$ such that execution time of an algorithm is $\geq c_1 g(n)$ and $\leq c_2 g(n)$ when $n \geq n_0$, then running time of the algorithm is stated as $\Theta(g(n))$. Note that $\Theta(n)$ is the time complexity for the above example since $O(n) = \Omega(n)$ for $f(n) = 4n + 3$.

Example 3.4.3 Let the running time of an algorithm be $3n^2 + 5n$. Then, $3n^2 + 5n \geq n^2$, $\forall n_0 \geq 1$, thus $\Omega(n^2)$. Let $3n^2 + 5n \leq c \cdot n^2$; this statement is true when $c = 8$ and $\forall n_0 \geq 1$, hence $O(n^2)$. Therefore, the running time of this algorithm is $\Theta(n^2)$ with $g(n) = n^2$, $c_1 = 1$, $c_2 = 8$ and $n_0 = 1$. The growth rate of these functions against the input size are depicted in Fig. 3.2.

3.5 Design Methods

Designing an algorithm to solve a particular problem clearly requires understanding the problem first. We can then apply one of the well-known design strategies that helps to find the solution effectively. The main methods of algorithm design are the *divide-and-conquer* method, the *greedy* method and *dynamic programming*.

3.5.1 Divide and Conquer

The divide-and-conquer method first breaks the problem into a number of subproblems which are smaller instances of the same problem. This method then recursively solves these subproblems and combines the results obtained to form the final output. This method commonly requires recursion which we will review in Chap. 6.

A simple example to illustrate this method is the summation of the elements of an array. We start by dividing the elements into small parts until each part has only two elements. We then start he addition with two elements and merge the results until the final sum is obtained as depicted in Fig. 3.3.

For this simple example, we did not need to apply divide-and-conquer strategy. However, we can apply the same procedure to sort the elements of an array which results in an efficient sorting algorithm called *mergesort*. The array is divided recursively into smaller segments until there are two elements remain and then sorting is carried from bottom to top in this algorithm.

3.5.2 Greedy Method

Greedy-algorithms aim to solve what is best with what is known at that point in time. These algorithms may find optimal solutions to some problems but may turn

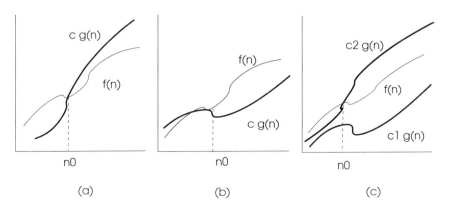

Fig. 3.2 Algorithm complexity classes, **a** worst case function $g(x)$, **b** best case function $g(x)$, **c** exact case function $g(x)$

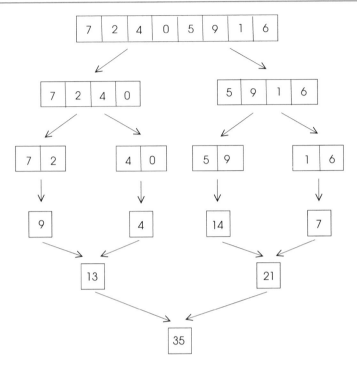

Fig. 3.3 Summation using divide and conquer

out to be far from optimal in various other problems. A greedy algorithm guesses the solution based on current local knowledge and iteratively repeats this procedure until a solution is reached.

Example 3.5.1 *Cashier's algorithm*: A cashier in a supermarket is frequently confronted with the task of paying a customer with the fewest number of coins. A greedy algorithm to accomplish this task is as follows. At each iteration, add the largest coin that does not pass the amount to be paid. For example, using the US coin system which has 1, 5, 10, 25, 100 cent coins; let the cashier pay \$2.78 cents in coins. The selection sequence with this algorithm will be 100, 100, 25, 25, 25, 1, 1, 1. Algorithm 3.14 shows the pseudocode of this algorithm.

Few things are to be noted in this procedure; it can return from two places, from line 8 when there is no solution and the last line when the correct sequence is found. It is possible to have no solutions when for example there existed 2 cents and not 1 cents and the change was required for 38 cents. We need to subtract the currently selected coin *largest* from the current amount not processed and the union symbol (∪) simply adds *largest* to the selected coin list.

The Cashier's Algorithm is optimal for the US coin system but it can be shown that it is sub-optimal, that is, may not find the result in shortest time for other systems. For example, US postal system has stamps 1, 10, 21, 34, 70, 100 cents etc. If we

Algorithm 3.14 *Cashier's Algorithm*

```
1: procedure CASHIER(change: integer)
2:     coins = {c₁, c₂, ..., cₙ}
3:     seq ← 0
4:     while change ≠ 0 do
5:         let largest be the largest coin such that largest ≤ change
6:         if largest = 0 then
7:             return "no solution"
8:         end if
9:         change ← change − largest
10:        seq ← seq ∪ largest
11:    end while
12:    return seq
13: end procedure
```

applied Cashier's Algorithm to buy stamps for 125 cents, the sequence will be 100, 21, 1, 1, 1, 1 with 6 steps. However, 70, 34, 21 cent sequence is shorter with 3 steps.

3.5.3 Dynamic Programming

This method is a powerful paradigm for algorithm design which dates back to 1950s. The problem is divided into subproblems first and assuming the optimal solution to a problem consists of optimal subproblems, the final optimal solution is formed. Different than the divided-and-conquer paradigm, the intermediate solutions are used to form improved solutions. Commonly, a table that consists of optimal solutions to subproblems is constructed. Then, these solutions are combined using procedures dependent on the problem that is under consideration. Using such a table avoids re-computation of intermediate results.

Example 3.5.2 Fibonacci sequence invented by the mathematician Fibonacci is given by $F = \{0, 1, 1, 2, 3, 5, 8, ...\}$ where each element of the sequence is the sum of its two preceding elements. We will write a procedure using a *for* loop that inputs n and outputs the nth element of this sequence. We can start by initializing the first two elements of the array to 0 and 1, and then gradually calculate the next element in this sequence. This algorithm is an example of dynamic programming since finding the nth element of this sequence is divided into subproblems that calculate all the preceding elements.

Algorithm 3.15 *Fibonacci Sequence*

```
1: procedure FIB(n: integer)
2:     F[n]: integer
3:     F[1] ← 0
4:     F[2] ← 1
5:     for i = 3 to n do
6:         F[i] ← F[i − 1] + F[i − 2]
7:     end for
8:     return F[n]
9: end procedure
```

3.6 Difficult Problems

We have stated that an algorithm should terminate after a certain number of steps. Moreover, the general requirement is that an algorithm should finish in some fore-seeable time with large input sizes. If we need to run an algorithm frequently, say few times a day with different input values and each output affecting the input of the next algorithm activation; then having this algorithm find the results in a few days will not be sensible. Some functions such as exponentials and factorial grow very fast with increased input size; for example, factorial of 6 is 720 and factorial of 10 is 3628800. Factorial of 100 is a huge number yet 100 is a modest input size for many problems. Having a running time of $n!$ for an algorithm with input size n means that the algorithm will not terminate in any foreseeable time for any significantly large n value.

These problems are very difficult to solve in general, and one way to tackle these problems is to concede to suboptimal solutions which work in reasonable time rather than optimal solutions that work in almost infinite time. These new class of algorithms are called *approximation algorithms* and they are used by many practical applications. An important merit of these algorithms is to determine how affinity of the result produced to the result obtained by the optimal algorithm. This closeness is termed *approximation ratio* which shows how effective the approximation algorithm is. In a large number of cases, approximation algorithms are not known to date and the general approach is to apply *heuristics* which are common sense rules. Heuristics cannot be proven to be correct and rigorous tests with different input patterns are needed to show they work fine for the problem at hand. Lastly, there exists problems that are known to be unsolvable. One such problem is the *halting problem* which is, given an arbitrary computer program, to determine whether the program will finish running or continue to run forever. It was proven by Alan Turing in 1936 that there can be no general algorithm to solve this problem, that is, to decide if a self-contained computer program will eventually halt [4]. Turing showed that an algorithm that correctly decides whether a program will halt or not can be made to contradict itself, thus it cannot be correct.

3.7   Review Questions

1. What are the main requirements from an algorithm?
2. What is the difference between an algorithm and a program?
3. Can a pseudocode of an algorithm work on a computer?
4. Write a three nested if structures with else statements identing each else with the if statement it belongs.
5. What is the main difference between a *for* loop and a *while* loop?
6. What is a procedure?
7. What is the difference between a procedure and a function?
8. What is the difference between call-by-value and call-by-reference?
9. What are the main methods of algorithm design?
10. What is the main difference between divide-and-conquer method and dynamic programming?
11. Give examples of a problems solved in polynomial time and exponential time.
12. What are the main approaches in search of a solution when confronted with a difficult problem that does not have a solution in polynomial time?

3.8   Chapter Notes

An algorithm is a finite sequence of instructions to solve a particular problem. It has 0 or more inputs, and produces outputs possibly working on the input. A program is the implementation of an algorithm in a way that can be easily understood by a computer. Unlike an algorithm, a program may not terminate, for example an operating system of a computer which waits in a loop is a program. A pseudocode is a representation of an algorithm using daily language and algorithm structures.

An algorithm uses assignment, decision and loop structures to solve a problem. Assignment is the process of assigning values to variables, decisions are taken to divert the flow of control during execution of the algorithm. Loops are mainly used to perform repeated running of a part of algorithm, possibly on different data. A procedure or a function of an algorithm is a sub-algorithm that may be used many times by calls.

The time and space complexities of an algorithm are the two basic merits displaying its effectiveness. The time complexity is the number of steps required by the algorithm in terms of the input size whereas the space complexity is the memory space needed by the algorithm during its running. The worst, best and average time complexity of an algorithm are commonly used to specify its performance.

Major algorithm design methods are the greedy, divide and conquer and dynamic programming approaches. The greedy method selects the best choice based on the current knowledge and works for only a small set of known problems. Divide and conquer method divides the problem into smaller parts, finds the solutions in smaller parts and then merges them. Dynamic programming also works on smaller instances of a problem but uses the intermediate outputs to construct larger outputs and it is one of the most powerful algorithm design methods.

Algorithms may be *parallel* in which case a number of parts of an algorithm work on different closely coupled computing hardware and cooperate to achieve a common goal. A distributed algorithm on the other hand refers to algorithms running on different computational nodes distributed over a network. A distributed algorithm communicates and synchronizes with algorithms on other nodes and solves a common problem over the network. A thorough description of the topics we summarized in this chapter can be found in [1–3].

Exercises

1. Write the pseudocode of simple algorithm that inputs an integer n and outputs n^2.
2. Write the pseudocode of an algorithm that inputs a positive integer n and outputs all odd integers from 1 up to and including n. For example, if 10 is input, the output is 1 3 5 7 9.
3. Write the pseudocode of a procedure that inputs a positive integer n and calculates the sum of integers from 1 to n and outputs the sum.
4. Write the pseudocode of an algorithm that inputs integers until a 0 is entered and calculates the sum of all even integers entered and displays it. For example, if 9 6 -2 3 4 8 0 is entered, the output is 16.
5. Write the pseudocode of an algorithm that inputs three integers a, b and c and outputs them in the order of magnitude from larger to smaller. For example, if -2 5 1 is entered, the output is 5 1 -2.
6. Write the pseudocode of an algorithm that inputs two integer arrays A and B, compares their corresponding element values and outputs a 1 if they are equal and a 0 if they are not.
7. Form a structure *employee* with a social security number (integer), age (integer) and wage (real number) and an array of employees named *factory*. Write the pseudocode of an algorithm that inputs the *factory* array and outputs all employee social security numbers whose age is greater than 30. The algorithm should also calculate the average age and extract wage of employees and output these values.
8. Work out the time complexity of an algorithm for $n \rightarrow \infty$ that runs in exactly $5n^2 + 6n + 7$ steps for an input size n.
9. Discuss the reasons of greedy algorithms failing to find the optimal solutions for most problems.
10. Describe the steps of the mergesort algorithms which sorts the elements of an integer array using divide-and-conquer method. Show the working of this algorithm in an array A={6,1,4,0,7,9,5,2}

References

1. Cormen TH, Leiserson CE, Rivest RL, Stein C (2009) Introduction to algorithms, 3rd edn. MIT Press, Cambridge
2. Dasgupta S, Papadimitriou, Vazirani U (2011) Algorithms. Sci Eng Math
3. Kleinberg J, Tardos E (2012) Algorithm design. Pearson, London
4. Turing, A (1937) On computable numbers, with an application to the Entscheidungsproblem, In: Proceedings of the London mathematical society, Series 2, vol 42, pp 230–265 (1937). https://doi.org/10.1112/plms/s2-42.1.230

Set Theory

4

Sets are fundamental structures in discrete mathematics. A set consists of elements that may or may not be related. One basic requirement when defining a set is that we should be able to decide whether a given object is an element of a set. For example, if set A consists of odd integers between 0 and 6, we can say 3 is an element and 4 is not an element of this set. A set can have a finite number of elements in which case it is denoted a *finite set* or it may consist of infinite number of elements in an *infinite set*. The finite set A of the example above has 3 elements as 1, 3 and 5. On the other hand, a set that contains all positive even numbers is infinite.

We describe fundamental concepts in set theory in this chapter starting with the definitions. We then continue with basic set operations and then with laws of set theory. We also show few simple algorithms to perform set operations.

4.1 Definitions

A *set* is an unordered collection of objects. The objects are called the *elements* or the *members* of the set. Uppercase letters are commonly used to represent the sets and lowercase letters denote the members. The elements of a set are shown within set braces (curly brackets). For example,

$$S = \{a, b, c, d\}$$

shows a set S consisting of four elements a, b, c and d. Elements of a set may be in any order; for example, the sets $\{a, c, b, d\}$ and $\{b, a, d, c\}$ are the same. Also, repetition of an element results in the same set, for example, $S = \{a, c, b, d\} = \{a, c, b, c, a, d\}$. In order to show that an element is a member of a set, we use $a \in S$ as in this example, and $e \notin S$ means e is not a member of the set S. A set can have related elements or it can consist of totally unrelated elements. As an example of the former, a set S that contains all positive integers less than 10 that are divided by 3

© Springer Nature Switzerland AG 2021

K. Erciyes, *Discrete Mathematics and Graph Theory*, Undergraduate Topics in Computer Science, https://doi.org/10.1007/978-3-030-61115-6_4

has 3, 6 and 9 as its elements. Note that we have defined a property of the members of the set S rather than explicitly specifying its elements. Using such a *set builder* notation, we can specify this set S as follows.

$$S = \{x | x \text{ is a positive integer less than } 10 \text{ and } x \pmod 3 = 0\}$$

which is an *implicit description* of a set whereas stating all members of a set is *explicit description* of the set. Implicit description is particularly useful when a set has many elements. Some specific sets of numbers that are worth noting are as follows.

$\mathbb{R}$, the set of real numbers
$\mathbb{R}^+$, the set of positive real numbers
$\mathbb{C}$, the set of complex numbers
$\mathbb{Q}$, the set of rational numbers
$\mathbb{Z} = \{..., -1, 0, 1, ...\}$, the set of integers
$\mathbb{Z}^+ = \{1, 2, 3, ...\}$, the set of positive integers
$\mathbb{N} = \{0, 1, 2, 3...\}$, the set of natural numbers

We can then use these sets to define a specific set as follows.

$$A = \{x \in \mathbb{Z}^+ | x \text{ is even and } x \leq 8\}$$

which means $A = \{2, 4, 6, 8\}$. The set with no elements is called the *empty set* and is shown by $\varnothing$. Note that $\{\varnothing\}$ is a set that is not empty and contains the empty set as its only element. An infinite set is shown to continue with '...' to show it progresses indefinitely, for example, $S = \{1, 3, 5, ...\}$ shows a set that has all positive odd numbers as its members. In some cases, we use '...' between the elements of a set with many elements and sometimes a set with unknown number of elements to avoid listing all members of the set. For example, $S = \{2, 4, 6, ..., 40\}$ is a set that has all positive even integers between and including 2 and 40 and $S = \{1, 3, 5, ..., n\}$ is the set of all positive odd integers up to n.

4.1.1 Equality of Sets

One of the first operations we may want to perform on two sets is two check whether they are equal or not.

Definition 4.1 (*Equality*) Two sets are equal if they have the same elements as their members. Formally, given two sets A and B, their equality means $\forall x, x \in A$ if and only if $\forall x \in B$, that is; $x \in A \leftrightarrow x \in B$.

The elements of two equal sets can be listed in any order as noted. For example, the sets $A = \{1, a, 8, x\}$ and $B = \{8, 1, x, a\}$ are equal. The inequality of two sets A and B is written as $A \neq B$ and can be assessed by taking the complement of the equality equation as follows,

$$\neg(\forall x : (x \in A \leftrightarrow x \in B)) \equiv \exists((x \in A \wedge x \notin B) \vee (x \notin A \wedge x \in B)$$

which is to say that there exists at least one element x that belongs to either set A or set B but not both.

4.1.2 Cardinality of a Set

We are often interested in the number of elements in a finite set. The size of a set is called its *cardinality*.

Definition 4.2 (*Cardinality of a set*) The *cardinality* of a finite set is the number of elements contained in it. The cardinality or the size of a set S is denoted by $|S|$ which can also be shown as $n(A)$ or $\#A$.

For example, given the following set,

$A = \{x \mid x$ is a positive integer between and including 2 and 12 and $x \pmod 4 = 0\}$

We can see the set A has a size of three consisting of elements $\{4, 8, 12\}$. Clearly, we can define cardinality of only a finite set.

4.2 Subsets

Another point of interest is to check whether a set is contained in another set or not.

Definition 4.3 (*Subset*) If every element of a set A is also an element of another set B, we say A is a *subset* of B and show this relation as $A \subseteq B$. Note that this property includes equality of the two sets. If set B contains one or more elements that are not members of the set A, then A is a *proper subset* of B and this is shown as $A \subset B$.

An element x of a set A is shown as $x \in A$ as noted before. Note that $\{x\} \subseteq A$ is different as it means the set $\{x\}$ is a subset of the set A. Using set builder notation, the subset relationship between two sets $A \subseteq B$ can be expressed as,

$$A \subseteq B = \{x \in U \mid x \in A \rightarrow x \in B\}$$

where U is the universal set. In other words, if an element x belongs to a set A, it must also belong to the set B for A to be a subset of B which means to prove $A \subseteq B$, we need to show that every element of set A is also an element of B. In order to prove that A is not a subset of B, we need to find at least one element of A that is not a member in set B. A set A that is not a subset of a set B is shown as $A \nsubseteq B$. We can then write the following using the rules of inversion of quantifiers.

$$A \nsubseteq B = \neg(A \subseteq B)$$
$$= \neg(\forall x : x \in A \rightarrow x \in B)$$
$$= \exists x : x \in A \wedge x \notin B$$

which means there exists at least one element that belongs to set A but not to B. We will now show an algorithm that tests whether a given set A of cardinality n is a proper subset, a subset or not of another set B of cardinality m as shown in Algorithm 4.1. We check whether each element of A is also an element of B and when this fails, the algorithm stops and not_subset output is returned. Otherwise elements of A are contained in B, but B should have extra elements for A to be denoted as a proper subset of B and this can be tested between lines 13 and 15. The best (minimum) execution time of this algorithm is when the first element of A is not found in B and this can be accomplished in unity time, and hence $\Omega(1)$. We are usually interested in the worst running time which occurs when we have checked each element of A with each element of B in nm time and therefore this is $O(nm)$ for this algorithm.

Algorithm 4.1 *Subset test*

1: **Input**: A, B ▷ input sets
2: **Output**: *not_subset*,*proper_subset* or *subset*
3: **int** $count \leftarrow 0$
4: **for all** $x \in A$ **do**
5:     **for all** $y \in B$ **do**
6:         **if** $x \neq y$ **then**
7:             **return** *not_subset*
8:         **else**
9:             $count \leftarrow count + 1$
10:        **end if**
11:    **end for**
12: **end for**
13: **if** $count < m$ **then**
14:     **return** *proper_subset*
15: **end if**
16: **return** *subset*

Subset relationship between two sets provide us with a method to prove that two sets are equal shown by the following theorem.

Theorem 1 *Given two sets A and B, if $A \subseteq B$ and $B \subseteq A$, then $A = B$.*

Proof If $A \subseteq B$, then $\forall x \in A$, $x \in B$ by the subset definition. Subsequently, If $B \subseteq A$, then $\forall x \in B$, $x \in A$ which shows equality.   □

Fig. 4.1 Subset relationship between sets $\mathbb{Z}$, $\mathbb{Q}$ and $\mathbb{R}$

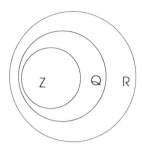

Definition 4.4 (*Power set*) The *power set* of a set A, denoted by $\mathcal{P}(A)$ is the set of all subsets of A including the empty set. The size of the power set of a set A is 2^n where n is the number of elements of A. Formally,

$$\mathcal{P}(A) = \{B \mid B \subseteq A\}$$

Example 4.2.1 Let us work out the power set of the set $A = \{a, b, c\}$. We can see the power set consists of the following sets, considering all combinations:

$$\mathcal{P}(A) = \{\{\varnothing\}, \{a\}, \{b\}, \{c\}, \{a, b\}, \{a, c\}, \{b, c\}, \{a, b, c\}\}$$

and it has $2^3 = 8$ elements since the set contains three elements. Note that the set containing the empty element is considered to exist as a subset of any set, that is, $\{\varnothing\} \subset A$ for any set A, and $\mathcal{P}(\varnothing) = \{\varnothing\}$ and $\mathcal{P}(\{\varnothing\}) = \{\varnothing, \{\varnothing\}\}$ ⊡

The set $\mathbb{Z}$ is a proper subset of the set $\mathbb{Q}$ since we can write any integer n as $n/1$ but $p/q \in \mathbb{Q}$ may not be an element in $\mathbb{Z}$, and $\mathbb{Q}$ is a proper subset of $\mathbb{R}$ but $\sqrt{2}$ for example is not a member of the set $\mathbb{Q}$ but a member of the set $\mathbb{R}$. This relationship is depicted in Fig. 4.1.

4.3 Venn Diagrams

Venn diagrams, invented by the mathematician John Venn, provide a graphical and hence a visual display of sets. The universal set U includes all of the objects that are considered and is shown by a rectangle. Circles or other shapes inside U represent the sets that are considered. Inside the circles and U, points are commonly used to denote the elements. Venn diagrams are useful in visualizing set relationships but they cannot be used to prove set equations.

Figure 4.2 displays the Venn diagram of three sets A, B and C enclosed in the universal set U; $A = \{x, y\}$, $B = \{z\}$, $C = \{u, w, z\}$ and $B \subset C$ as all elements of B (only z in this case) are also elements of C. The set A does not have any common elements with either of sets B and C.

Fig. 4.2 A Venn Diagram of three sets inside the universal set

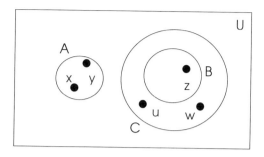

Fig. 4.3 The Venn Diagram of the complement A^c of the set A is shown in grey

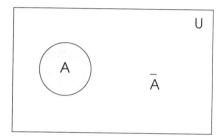

4.4 Set Operations

Operations on sets commonly involve generating new sets from the input sets. Basic set operations are the complement, set union, intersection, difference and product operations described below.

Definition 4.5 (*Complement*) Given the universal set U, the *complement* A^c, commonly shown as $\overline{A}$, of a set in U is the set consisting of all of the elements in U that are not elements of A. Formally,

$$\overline{A} = \{x \in U | x \in \overline{A} \longleftrightarrow x \notin A\}$$

Given U as $\mathbb{N}$, let set A consist of all even integers in U. The complement of A, $\overline{A}$ is then all odd positive integers. The Venn diagram of the complement of a set is illustrated in Fig. 4.3.

Definition 4.6 (*Union*) The *union* C of two sets A and B, shown as $A \cup B$ is the set consisting of all of the elements in A and B. Formally,

$$A \cup B = \{x \in U | (x \in A \cup B) \longleftrightarrow (x \in A) \vee (x \in B)\}$$

Given two sets $A = \{a, b, c, d, e\}$ and $B = \{b, c, d, e, f, g\}$, their union $A \cup B$ is $\{a, b, c, d, e, f, g\}$ as shown in Fig. 4.4.

The union of more than two sets is a set consisting of all elements in these sets. Given sets $A_1, ..., A_n$, their union set A can be stated formally as follows.

Fig. 4.4 The Venn Diagram
of the union of two sets A
and B shown in grey

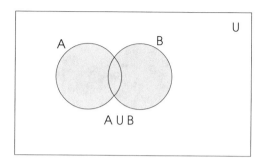

$$A = \bigcup_{i=1}^{n} A_i = A_1 \cup A_2 \cup ... \cup A_n$$

An algorithm to find the union of two sets is shown in Algorithm 4.2. The output set C is initialized to have no elements first. All of the elements of the set A is included in the union set C then and any member of set B that is not already contained in set C is then made member of C to prevent inserting the set element that are in intersection of A and B twice. The time complexity of this algorithm for $|A| = n$ and $|B| = m$ is $\Theta(nm)$, considering the first loop runs n times and the second loop runs nm times since testing $x \in C$ has to be done with every element of the set C for a total of m times.

Algorithm 4.2 *Set union*

1: **Input**: A, B ▷ input sets
2: **Output**: C ▷ union of A and B
3: **int** $i, j, count$
4: $C \leftarrow \emptyset$
5: **for all** $x \in A$ **do**
6:     $C \leftarrow C \cup \{x\}$
7: **end for**
8: **for all** $x \in B$ **do**
9:     **if** $x \notin C$ **then**
10:        $C \leftarrow C \cup \{x\}$
11:    **end if**
12: **end for**

Definition 4.7 (*Intersection*) The *intersection* C of two sets A and B is the set consisting of all of the elements in both A and B. Formally,

$$A \cap B = \{x \in U \,|\, x \in (A \cap B) \longleftrightarrow (x \in A) \wedge (x \in B)\}$$

Fig. 4.5 Intersection of two
sets A and B. The elements
u and t belong to $A \cap B$

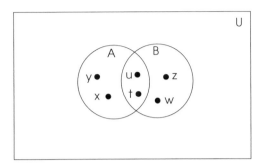

Given two sets $A = \{a, b, c, d, e\}$ and $B = \{b, c, d, e, f, g\}$, their intersection $A \cap B$ is $\{b, c, d, e\}$. The Venn diagrams for the intersection of two sets A and B is shown in Fig. 4.5.

The algorithm shown in Algorithm 4.3 displays how the intersection of two sets can be formed. Each element of one set is searched in the other. Two nested for loops are executed nm times giving $\Theta(nm)$ time complexity for this algorithm.

Algorithm 4.3 *Set intersection*

1: **Input**: A, B ▷ input sets
2: **Output**: C ▷ intersection set
3: $C \leftarrow \emptyset$
4: **for all** $x \in A$ **do**
5:     **for all** $y \in B$ **do**
6:         **if** $x = y$ **then**
7:             $C \leftarrow C \cup \{x\}$
8:         **end if**
9:     **end for**
10: **end for**

The cardinality of a unions of sets A and is,

$$n(A \cup B) = n(A) + n(B) - n(A \cap B)$$

because when counting the members of the union of sets A and B, we count the members in A and members in B and add them, and since we add the elements that belong to two sets twice, we need to subtract the number of elements in the intersection of two sets.

Definition 4.8 (*Disjoint sets*) Two sets are said to be *disjoint* if their intersection is the empty set, that is, $A \cap B = \emptyset$. The intersection of two disjoint sets has no elements; formally, $|A \cap B| = 0$.

For example, given two sets $A = \{a, b, c\}$ and $B = \{1, 3, 8\}$, we can conclude they are disjoint as they have no common elements.

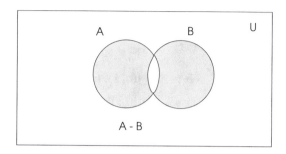

Fig. 4.6 Symmetric difference of two sets A and B shown in grey

Definition 4.9 (*Difference*) The *difference* set C of two sets A and B denoted by $A - B$ or $A \setminus B$ is the set consisting of all of the elements in A that are not elements of B. Formally,

$$A - B = \{x \in U | x \in A - B \longleftrightarrow (x \in A) \wedge (x \notin B)\}$$

Given two sets $A = \{a, b, c, d, e\}$ and $B = \{b, d, e\}$, their difference $A - B$ is $\{a, c\}$. Note that the complement of a set A can be written as $\overline{A} = U \setminus A$ where U is some universal set in which A is defined.

Definition 4.10 (*Symmetric difference*) The *symmetric difference* of two sets A and B, denoted by $A \Delta B$ is the set consisting of all of the elements that are either in A or B but not in both. Formally,

$$A \Delta B = \{x \in U \,| \, x \in A \Delta B \longleftrightarrow x \in A \oplus x \in B\}$$

The symmetric difference can also be expressed as,

$$A \Delta B = (A - B) \cup (B - A)$$

which is to say that this is the union of the elements of A that are not in B, and the elements of B that are not in A. For example, given two sets $A = \{a, b, c, d, e\}$ and $B = \{b, c, d, e, f, g\}$, their symmetric difference $A \Delta B$ is $\{a, f, g\}$. The symmetric difference of two sets A and B is shown in the Venn diagram of Fig. 4.6.

4.4.1 Cartesian Product

Definition 4.11 (*Cartesian product*) The cartesian product C of two sets A and B is the set of all ordered pairs (a, b) where $a \in A$ and $b \in B$. This product is denoted by $C = A \times B$.

Example 4.4.1 Let $A = \{1, 2, 3\}$ and $B = \{a, b\}$. Then,

$$A \times B = \{(1, a), (1, b), (2, a), (2, b), (3, a), (3, b)\}$$

$\square$

Note that if sets A and B have n and m elements, the cartesian product set C has nm elements. Also, this operation is not symmetric, that is, $A \times B \neq B \times A$. For the above example,

$$B \times A = \{(a, 1), (a, 2), (a, 3), (b, 1), (b, 2), (b, 3)\}$$

but the resulting set still has the same number of elements. This notion can be extended to a number of sets as follows.

Example 4.4.2 Let $A = \{1, 2\}$ and $B = \{a, b, c\}$ and $C = \{x, y\}$. Then,

$$A \times B \times C = \{(1, a, x), (1, a, y), (1, b, x), (1, b, y), (1, c, x), (1, c, y), (2, a, x),$$
$$(2, a, y), (2, b, x), (2, b, y), (2, c, x), (2, c, y)\}$$

This time, the product has k elements where k is the product of the cardinalities of the sets.

We can have an algorithm to find the cartesian product of two sets as shown in Algorithm 4.4. We simply enlarge the cartesian product C of the sets A and B by combining an element from each set in each iteration. Time complexity is $O(nm)$ for $|A| = n$ and $|B| = m$ due to nested *for* loops.

Algorithm 4.4 *Cartesian product*

1: **Input**: A, B ▷ input sets
2: **Output**: C ▷ Cartesian product of A and B
3: $C \leftarrow \varnothing$
4: **for all** $x \in A$ **do**
5:     **for all** $y \in B$ **do**
6:         $C \leftarrow C \cup \{x, y\}$
7:     **end for**
8: **end for**

4.4.2 Set Partition

A set may be divided into a number of partitions defined below.

Definition 4.12 (*Set partition*) Consider a set A which consist of a number of nonempty subsets such that $A = A_1 \cup A_2 \cup ... \cup A_n$, and $A_1, A_2, ..., A_n$ are mutually disjoint with no common elements between any two pairs of these subsets, the collection of these subsets is called a *partition* of the set A.

Fig. 4.7 A partition of a set S

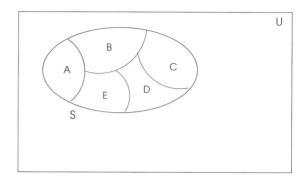

A partition of a set S that consists of disjoint subsets A, B, C, D and E is depicted in the Venn diagram of Fig. 4.7. For example, $\{\{a\}, \{b, c\}, \{d, e, f\}\}$ is a partition of the set $\{a, b, c, d, e, f\}$. Note that given $A = \mathbb{N}$, the sets $\{0\}$ and $\mathbb{Z}^+$ form a partition of A.

Example 4.4.3 Given $S_1 = \{a \in \mathbb{Z} | a = 3k\}$, for some integer k, $S_2 = \{a \in \mathbb{Z} | a = 3k + 1\}$, for some integer k and $S_3 = \{a \in \mathbb{Z} | a = 3k + 2\}$, for some integer k is a partition of $\mathbb{Z}$ since $\mathbb{Z} = S_1 \cup S_2 \cup S_3$. For $k = 0, 1, ..., S_1 = \{0, 3, 6, ...\}$, $S_2 = \{1, 4, 7, ...\}$, $S_3 = \{2, 5, 8, ...\}$ □

4.4.3   Operation Precedence

The precedence between the general set operations is the complement followed by set intersection and then set union. Given,

$$A \cup B \cap \overline{C}$$

is equal to

$$A \cup (B \cap (\overline{C}))$$

4.5   Laws of Set Theory

The laws of set theory follow a similar pattern to the laws of logic as listed below. Most of these laws are common sense as were the logic laws. Note that De Morgan's laws have a similar structure but are defined for sets this time.

1. *Identity Laws*:
$$A \cup \varnothing = A$$

Union of a set A with the empty set is itself.

$$A \cap U = A$$

Intersection of a set A with the universal set is itself.

2. *Idempotent Laws*:

$$A \cup A = A$$

$$A \cap A = A$$

3. *Inverse laws*:

$$A \cup \overline{A} = U$$

$$A \cap \overline{A} = \varnothing$$

4. *Domination laws*:

$$A \cap \varnothing = \varnothing$$

A set A has no common elements with the empty set.

$$A \cup U = U$$

Union of a set A with the universal set is the universal set.

5. *Commutative laws*:

$$A \cup B = B \cup A$$

$$A \cap B = B \cap A$$

6. *Absorption laws*:

$$A \cup (A \cap B) = A$$

Any element in the intersection of the sets A and B is a member of the set A and hence union of such elements with A will be A.

$$A \cap (A \cup B) = A$$

Intersection of set A with its union with another set is the set A.

7. *Associative laws*:

$$A \cup (B \cup C) = (A \cup B) \cup C$$

$$A \cap (B \cap C) = (A \cap B) \cap C$$

8. *Distributive laws*:

$$A \cup (B \cap C) = (A \cup B) \cap (A \cup C)$$

$$A \cap (B \cup C) = (A \cap B) \cup (A \cap C)$$

9. *De Morgan's laws*:

$$\overline{A \cup B} = \overline{A} \cap \overline{B}$$

$$\overline{A \cap B} = \overline{A} \cup \overline{B}$$

4.6 Proving Set Equations

We can have various methods to prove equations involving sets. We will elaborate on three such ways which are the *element method*, the *tabular method* and *proofs with quantifiers*. A Venn diagram is illustrative in showing the results of a proof but is not considered as a formal proof method as noted.

4.6.1 The Element Method

In this method, we commonly start with an element x belonging to the left side of the equation and using the set properties, we attempt to arrive at a known rule at the right side of the equation. Two proof cases to apply this method are as follows.

- *Proving subset inequalities*: An arbitrary but a distinct element in the left side is considered and we try to show that $(x \in L) \rightarrow (x \in R)$ where L and R are the left and right sides of the equation. Note that proving this ensures $L \subseteq R$ since subset relationship is defined this way.

Theorem 2 $A \subseteq B$ and $A \cap \overline{B} = \varnothing$ are equivalent.

Proof We will prove first part of this theorem using the contrapositive argument, that is, if $(A \cap \overline{B} \neq \varnothing)$ then $(A \nsubseteq B)$. Since $(A \cap \overline{B} \neq \varnothing)$, $\exists x$ such that $x \in A$ and $x \in \overline{B}$ which means $x \notin B$. That is,

$$x \in A \wedge x \notin B$$

which means $(A \nsubseteq B)$ by the definition of a subset.

$\square$

- *Proving equations*: This time, we need to show $(x \in L) \rightarrow (x \in R)$ and $(x \in R) \rightarrow (x \in L)$ for the equality to hold $(L \leftrightarrow R)$.

Example 4.6.1 We will prove De Morgan's second law which states $\overline{A \cap B} = \overline{A} \cup \overline{B}$. What we should prove here is that if x is an element of the inverse of the intersection of two sets A and B, it should not be contained in the intersection of these two sets, therefore it is a member of the union of these sets.

Proof

$$
\begin{aligned}
x \in \overline{A \cap B} &\rightarrow x \notin (A \cap B) \\
&\rightarrow (x \in A) \vee (x \in B) \\
&\rightarrow x \in (A \cup B)
\end{aligned}
$$

In the other direction,

$$x \in \overline{(\overline{A} \cup \overline{B})} = \overline{B} \to x \notin (\overline{A} \cup \overline{B})$$
$$= \overline{B} \to x \notin (A \cap B)$$
$$= \overline{B} \to x \in \overline{A \cap B}$$

□

Example 4.6.2 Prove $A \cap \overline{B} = A - B$.

Proof We start by assuming a general member x in the left side of the equation as follows.

$$x \in A \cap \overline{B} \to x \in A \wedge x \in \overline{B}$$
$$\to x \in A \wedge x \notin B$$

and this is the exact definition of $A - B$ in Definition 4.9 and hence any $x \in L$ is also a member of the right side ($x \in R$). Next, we consider $x \in A - B$.

$$(x \in A - B) \to (x \in A) \wedge (x \notin B)$$

which means x is a member in $\overline{B}$,

$$\to (x \in A) \wedge (x \in \overline{B})$$

By the definition of intersection, we can write,

$$\to (x \in A \cap \overline{B})$$

We have shown ($x \in R \to x \in L$) which completes the proof.   □

This equality is depicted in Fig. 4.8 with the aid of a Venn diagram where (a) shows two sets A and B with their intersection, (b) displays set B and $\overline{B}$ only, discarding set A, and difference $A - B$ is shown in (c).

Example 4.6.3 Prove $A - (A - B) = A \cap B$.

Proof Consider $x \in L$,

$$(x \in (A - (A - B))) \to (x \in A) \wedge (x \notin (A - B)$$
$$\to (x \notin A) \vee (x \in B) \text{ since } x \in A , \text{ specialization yields,}$$
$$\to (x \in B)$$
$$\to (x \in A) \wedge (x \in B)$$
$$\to (x \in A \cap B)$$

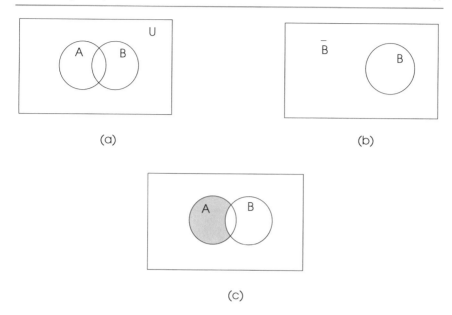

Fig. 4.8 The Venn Diagram of $A - B$ shown in grey

Now, we need to prove the other direction,

$$(x \in (A \cap B)) \rightarrow (x \in A) \wedge (x \in B)$$
$$\rightarrow x \notin (A - B) \text{ since } x \in A \text{ and by definition of set difference,}$$
$$\rightarrow x \in A - (A - B)$$

$\square$

Example 4.6.4 Prove that $A \cup (B \cap C) = (A \cup B) \cap (A \cap C)$.

Proof We will prove first prove that when left side of the equation holds, the right side also holds. We have two cases for the left side of the equation, either $x \in A$ or $x \in B \cap C$.

- Case 1: $x \in A$, then,

  $x \in A \cup B$ by the definition of union
  $x \in A \cup C$ by the definition of union
  $\therefore x \in (A \cup B) \cap (A \cup C)$ by the definition of intersection
- Case 2: $x \in (B \cap C)$, then,

  $(x \in B) \wedge (x \in C)$ by the definition of intersection
  $(x \in A \cup B) \wedge (x \in A \cup C)$ by the definition of union
  $\therefore x \in (A \cup B) \cap (A \cup C)$ by the definition of intersection

Now let us consider proving the equation from right to left for two cases when $x \in A$ and $x \notin A$.

- Case 1: $x \in A$, then
$$x \in A \cup (B \cap C) \text{ by the definition of union}$$

- Case 2: $x \notin A$, then
$$(x \notin A) \wedge (x \in (A \cup B)) \quad \therefore x \in B$$
$$(x \notin A) \wedge (x \in (A \cup C)) \quad \therefore x \in C$$
$$\therefore x \in A \cup (B \cap C) \text{ by the definition of intersection}$$

$\square$

which proves the reverse direction.

Example 4.6.5 Prove for all sets A, B and C, if $A \subseteq B$ and $B \subseteq \overline{C}$, then $A \cap C = \varnothing$.

Proof We will prove this proposition using the contradiction method combined with the element method. Assume,
$$\exists x \in (A \cap C)$$
by definition of intersection,
$$(x \in A) \wedge (x \in C)$$
$(x \in A)$ and $(A \subseteq B)$, then
$$x \in B$$

$$(x \in B) \wedge (B \subseteq \overline{C}) \rightarrow x \in \overline{C}$$
therefore $x \notin C$ which contradicts the second statement. $\square$

4.6.2   The Tabular Method

In this case, we build the *membership table* for each side of the equation, taking each set as input variables and listing all of their possible values, similar to what we have done for compound propositions. A true value (1) in a row is interpreted as x is a member; for example, a 1 under the set A means $x \in A$. When $x \in A \wedge x \in B$, then we place a 1 under $A \cap B$.

Example 4.6.6 Prove De Morgan's first law associated with sets, which is $\overline{A \cup B \cup C}$ $= \overline{A} \cap \overline{B} \cap \overline{C}$ using the tabular method.

Proof We form the truth table for the both sides of the equation as shown in Table 4.1. The last two columns of the table are respectively the left side and the right side of the equation and they are equal. $\square$

Table 4.1 Truth table for De Morgan's first set law

A	B	C	$\overline{A}$	$\overline{B}$	$\overline{C}$	$A \cup B \cup C$	$\overline{A \cup B \cup C}$	$\overline{A} \cap \overline{B} \cap \overline{C}$
0	0	0	1	1	1	0	1	1
0	0	1	1	1	0	1	0	0
0	1	0	1	0	1	1	0	0
0	1	1	1	0	0	1	0	0
1	0	0	0	1	1	1	0	0
1	0	1	0	1	0	1	0	0
1	1	0	0	0	1	1	0	0
1	1	1	0	0	0	1	0	0

4.6.3 The Algebraic Method

The algebraic method employs laws of set theory to prove set equations and inequalities.

Example 4.6.7 Prove $(A \cup B) - C = (A - C) \cup (B - C)$

Proof

$$(A \cup B) - C = (A \cup B) \cap \overline{C} \text{ set identity law}$$
$$= \overline{C} \cap (A \cup B)$$

De Morgan's law

$$= (\overline{C} \cap A) \cup (\overline{C} \cap B)$$
$$= (A \cap \overline{C}) \cup (B \cap \overline{C})$$
$$= (A - C) \cup (B - C)$$

□

Theorem 3 *Prove that $A \cap \overline{B} = \varnothing$ and $\overline{A} \cup B = U$ are equivalent.*

Proof Assume $A \cap \overline{B} = \varnothing$, then,

$$\overline{A \cap \overline{B}} = \overline{\varnothing}$$

Using De Morgan's laws,

$$\overline{A} \cup \overline{\overline{B}} = U$$
$$\overline{A} \cup B = U$$

□

Therefore, $A \subseteq B$, $A \cap \overline{B} = \varnothing$ and $\overline{A} \cup B = U$ are equivalent. The equivalence of the first two statements was proved in Theorem 2.

4.7 Review Questions

1. Compare a subset and a proper subset of a set.
2. How can two sets be proven to be equal using the subset concept?
3. What is meant by the power set of a set?
4. What is the magnitude of the power set of a set with k elements?
5. Can a Venn diagram be used to prove a set equation?
6. Describe the union, intersection and cartesian product of two sets.
7. What is the symmetric difference of two sets A and B?
8. How is the partition of a set defined?
9. What are the main methods of proving set equations?

4.8 Chapter Notes

Sets are one of the main topics of study in discrete mathematics. We have reviewed definitions and the main properties, basic concepts related to sets. One such property between two sets is whether they are equal or one is contained in the other. The subset relation $\subseteq$ between two sets A and B means all members of A are contained in B. Moreover this relation can be used to prove two sets are equal, that is, if $A \subseteq B$ and $B \subseteq A$, then $A = B$. We can use this property to prove various set equations. Venn diagrams are used to visualize set relations. Various operations on sets such as union, intersection, cartesian product can be defined. We provided algorithms that input two sets and perform these operations.

Laws of set theory provide a solid basis to prove set equations. We have the element method, tabular method and the algebraic method to prove to prove set equations. The element method selects an arbitrary element x from the left side of the equation and we need to show $(x \in L) \rightarrow (x \in R)$ and $(x \in R) \rightarrow (x \in L)$ where L is the left side and R is the right side of the equation, for the equality to hold. The tabular method is used to prove set equations by listing all the possibilities of the left hand side and the right hand side of the equation and then checking whether all rows are equal. The algebraic method uses laws of set theory to prove the equalities. Set theory is reviewed in [1, 2] in detail.

Exercises

1. Given $A = \{3, 9, f, T\}$ and $B = \{9, 3, T, f\}$, is $A = B$?
2. Sets $A = \{2, 0, 1, 12\}$ and $B = \{1, 12, 0, 5, 7\}$ are given. Is $A \subset B$?
3. What set is the union of sets $A = \{5, 10, 3, 7\}$ and $B = \{7, 3, 8, 9, 5\}$
4. What is the intersection of sets $A = \{2, 0, 1, 12\}$ and $B = \{1, 12, 0, 5, 7\}$
5. Given $A = \{12, a, 3, 5, 7, 11\}$ and $B = \{3, 6, b, 9, 11, 8\}$, what is $A - B$?
6. Let $U = \{x : 1 \leq x < 30\}$ as the universal set, O be the odd numbers in U, P be the prime numbers in U and $A = \{1, 2, 6, 7, 9, 13, 17\}$, find the following.

a. $\overline{O}$

c. $P \cap O$

b. $O \cap A$

d. $O \cup A \cap \overline{P}$

7. Sets $A = \{3, 1, 7, 9, 8, 5\}$ and $B = \{2, 9, 1, 5, 6\}$ are given. Determine $\mathcal{P}(A)$, $\mathcal{P}(B)$, $\mathcal{P}(A \cup B)$, $\mathcal{P}(A \cap B)$ and $\mathcal{P}(A \times B)$.
8. Given two sets A and B with $A \cap B \neq \emptyset$, draw the Venn diagrams for the following sets.

a. $B - A$

c. $\overline{(A \cup B)}$

b. $A \Delta B$

d. $\overline{A} \cap \overline{B}$

9. Let sets $A = \{a, b, c\}$, $B = \{3, 4\}$ and $C = \{x, y, x\}$. Find $A \times B \times C$.
10. Given sets A, B and C, prove that if $A \not\subseteq B$ and $B \not\subseteq C$, then $A \not\subseteq C$.
11. Prove De Morgan's first law $\overline{A \cup B} = \overline{A} \cap \overline{B}$ using membership method.
12. Prove that $(A \cap B) \cap C = A \cap (B \cap C)$ using tabular method.
13. Prove the following equalities.

a. $A \cup (A \cap B) = A$

c. $(A \cap B) \cup (A \cap \overline{B}) = A$

b. $A \cap (A \cup B) = A$

d. $B \setminus A = B \cap \overline{A}$

14. Write the pseudocode of an algorithm that inputs two sets A and B and tests whether they are disjoint.

References

1. Eppstein SS (2010) Discrete mathematics with applications. Cengage learning, 4th edn
2. Rosen KH (2011) Discrete mathematics and its applications, 7th edn. McGraw-Hill Education, New York

Relations and Functions

5

A set of objects may be related to another set of objects with some similarity. A relation associates an element of a set with an element of another set. Let the set A be consisting of names of persons as David, Jose, Kemal and the set B be the some cities as Istanbul, London and Madrid. A relation R can then be defined as the city in set B where a person in set A lives. For example, (David, London), (Kemal, Istanbul) and (Jose, Madrid) can all be the elements of the relation.

In the first part of this chapter, we review basic definitions related to relations, representation of operations on relations and types of relations. A function is a special relation that relates one or more elements of a set to exactly one element of another set. In the second part, we describe functions and their properties.

5.1 Relations

Given two non-empty, finite sets A and B, a (binary) relation R is a set of ordered elements (a, b) with $a \in A$ and $b \in B$. The element a is said to be related to b through R, or $(a, b) \in R$, or $a R b$ to mean the same thing. A binary relation R from a set A to a set B can be viewed as a mapping from the cartesian product $A \times B$ to the set $\{0, 1\}$ where the 1 value in $\{0, 1\}$ for a pair (a, b) with $a \in A$ and $b \in B$ means $a R b$.

Definition 5.1 (*binary relation*) Let A and B be two sets. A *binary relation* or simply a *relation* R from A to B (or between A and B) is a subset of the Cartesian product $A \times B$.

© Springer Nature Switzerland AG 2021

K. Erciyes, *Discrete Mathematics and Graph Theory*, Undergraduate Topics in Computer Science, https://doi.org/10.1007/978-3-030-61115-6_5

Fig. 5.1 Set diagram
representation of a relation

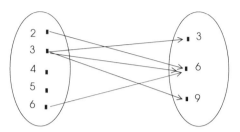

Example 5.1.1 Let $A = \{2, 3, 4, 5, 6\}$ and $B = \{3, 6, 9\}$ and relation $R = \{(a, b) : a$ divides $b\}$. Thus, (2,6), (3,3), (3,6), (3,9), (6,6) are the elements of R.

A relation may be defined between the elements of a single set as in the example below. In this case, we consider each element of set A as $B = A$. This means a relation on a set A is a subset of $A \times A$, and such a relation is called a *unary relation*. Commonly, a general set such as $\mathbb{N}$ or $\mathbb{Z}$ is given and a relation is defined between the elements of this given set.

Example 5.1.2 Let $A = \{1, 2, 3, 4, 5, 6\}$ and and relation $R = \{(a, b) : b = 2a\}$ is given on set A. We can see (1,2), (2,4) and (3,6) are the elements of R.

5.1.1   Representations

We can have graphical or binary matrix representation of a relation.

5.1.1.1   Graphical Representation

The relation of Example 5.1.1 can be represented as in Fig. 5.1 in which sets A and B are represented by a diagram and a directed line is drawn between a and b if $(a, b) \in R$ for the relation stated in this example.

When $B = A$ for the same problem, we can draw the graph shown in Fig. 5.2 where each member of the set is shown by a circle and a directed line is drawn between the elements a and b if $(a, b) \in R$. This representation is called *directed graph (digraph) representation* of a relation.

5.1.1.2   Binary Matrix Representation

A *matrix* is a collection of numbers arranged in a fixed number of rows and columns. Each row and column of a matrix is numbered in increasing integers as $1, ..., n$. An element of a matrix is specified using its row and column. For example, given a matrix M, $M[2, 3]$ is the element at 2nd row and 3rd column. A *binary matrix*, which consists of elements with values 0 or 1 only, can be used to represent a relation

Fig. 5.2 Digraph
representation of the relation
of Example 5.1.1. When
$B = A$

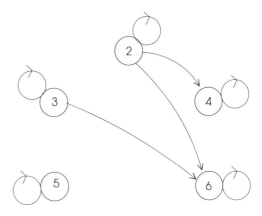

R from a set A to set B. In this case, the rows of the matrix are labeled with the
elements of set A and the columns show the elements of set B. There is a 1 in the
corresponding (a, b) pair with $a \in A$ and $b \in B$ if and only if $(a, b) \in R$. The
binary matrix representation of the relation of Example 5.1.1 is shown below. Using
a binary matrix to depict a relation is convenient when operations between two or
more relations are performed as we will see.

$$
\begin{array}{c c c c}
 & 3 & 6 & 9 \\
2 & \begin{pmatrix} 0 & 1 & 0 \\ 1 & 1 & 1 \\ 0 & 0 & 0 \\ 0 & 0 & 0 \\ 0 & 1 & 0 \end{pmatrix} \\
3 \\
4 \\
5 \\
6
\end{array}
$$

5.1.2 Inverse of a Relation

The inverse of a relation R, denoted by R^{-1} relates an element b to a for all elements
(a, b) of R.

Definition 5.2 (*inverse*) Let R be a relation from a set A to a set B. The inverse
relation R^{-1} is defined as follows:

$$R^{-1} = \{(b, a) \in (B \times A) | (a, b) \in R$$

Example 5.1.3 Let $A = \{2, 3, 5\}$, $B = \{1, 3, 8\}$ and R be $a \leq b$ with $a \in A$ and
$b \in B$. Find the elements of R and R^{-1} and draw the graphical representation of R
and R^{-1}.
Solution: $R = \{(2, 3), (2, 8), (3, 3), (3, 8), (5, 8)\}$ and $R^{-1} = \{(3, 2), (3, 3), (8, 2),$
$(8, 3), (8, 5)\}$. The graphs of R and R^{-1} are shown in Fig. 5.3 respectively. Note that
R^{-1} can be obtained simply by reversing the direction of arrows in R.

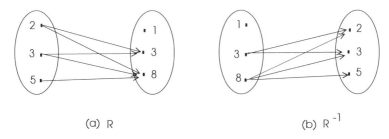

Fig. 5.3 Relation R and its inverse R^{-1}

Let us sketch a procedure to form the inverse R^{-1} of a relation R which consists of pairs (a_i, b_i) for $i = 1$ to n, input to the algorithm as shown in Algorithm 5.1. Any element $(a, b) \in R$ is copied as (b_i, a_i) to R^{-1}.

Algorithm 5.1 *Inverse of a Relation*

1: **procedure** INVERSE(R: Relation)
2:     $R^{-1} \leftarrow \varnothing$
3:     **for all** $(a, b) \in R$ **do**
4:         $R^{-1} \leftarrow R^{-1} \cup (b, a)$
5:     **end for**
6:     **return** R^{-1}
7: **end procedure**

5.1.3 Union and Intersection of Relations

Relations can be united or intersected as in the union and intersection of sets. The union of two relations R_1 and R_2 consists of all pairs of elements that are either in R_1, R_2 or both. The intersection of two relations R_1 and R_2 has elements that appear both in R_1 and R_2 only.

Example 5.1.4 Assume the following relations on the set $A = \{1, 2, 3, 4, 5\}$.

$$R_1 = \{(1, 2), (1, 5), (2, 4), (3, 5), (4, 4), (5, 1)\}$$
$$R_2 = \{(1, 2), (2, 4), (3, 3), (4, 4), (4, 2), (5, 1)$$

Then the union and intersection of these sets are as follows.

$$R_1 \cup R_2 = \{(1, 2), (1, 5), (2, 4), (3, 3), (3, 5), (4, 4), (4, 2), (5, 1)\}$$
$$R_1 \cap R_2 = \{(1, 2), (2, 4), (4, 4), (5, 1)\}$$

Algorithm 5.2 shows how to form the union of two sets R_1 and R_2 which consist of pairs (a_i, b_i) for $i = 1$ to n, and (c_i, d_i) for $i = 1$ to m respectively. Initially, the union R_T of R_1 and R_2 contains all elements of R_1 and any different element of R_2 than R_1 is copied to R_T. This way, duplicate copying is prevented.

Algorithm 5.2 *Union of Two Relations*

1: **procedure** UNION(R_1, R_2: Relation)
2:   $R_T \leftarrow R_1$
3:   **for** $i = 1$ to n **do**
4:     **for** $j = 1$ to m **do**
5:       **if** $(a_i, b_i) \in R_1 \neq (c_j, d_j) \in R_2$ **then**
6:         $R_T \leftarrow R_T \cup (c_j, d_j)$
7:       **end if**
8:     **end for**
9:   **end for**
10:   **return** R_T
11: **end procedure**

Finding the intersection of two relations R_1 and R_2 is depicted in Algorithm 5.3 where any common element is copied to the intersection set R_I.

Algorithm 5.3 *Intersection of Two Relations*

1: **procedure** INTERSECTION(R_1, R_2: Relation)
2:   $R_I \leftarrow \emptyset$
3:   **for** $i = 1$ to n **do**
4:     **for** $j = 1$ to m **do**
5:       **if** $(a_i, b_i) \in R_1 = (c_j, d_j) \in R_2$ **then**
6:         $R_I \leftarrow R_I \cup (a_i, b_i)$
7:       **end if**
8:     **end for**
9:   **end for**
10:   **return** R_I
11: **end procedure**

5.1.4 Properties of Relations

A relation on a set may be reflexive, symmetric, antisymmetric and transitive.

5.1.4.1 Reflexive Relation

A reflexive relation on a set relates each element to itself.

Definition 5.3 (*reflexive*) A relation R is reflexive if and only if $\forall a \in R$, $(a, a) \in R$ (or aRa). A relation R is not reflexive if $\exists a \in R$ such that $(a, a) \notin R$

Example 5.1.5 A relation R is defined in a set A of integers as $(a, b) \in R$ if and only if $a \le b$. R is reflexive because for each element $a \in A$, $(a, a) \in R$ because of equality.

Example 5.1.6 A relation R is defined in the set $\mathbb{Z}$ as $(a, b) \in R$ if and only if $(a + 4b)$ is divisible by 5. Find if R is reflexive or not.
Solution: Let $a \in \mathbb{Z}$. Then, $a + 4a = 5a$ which is divisible by 5. Therefore, aRa, $\forall a \in \mathbb{Z}$ and thus, R is reflexive.

Let us sketch a procedure to test whether a relation R is reflexive or not. We have R which consists of pairs (a_i, b_i) for $i = 1$ to n, input to the algorithm and we test whether for any $(a_i, b_i) \in R$, whether $(a_i, a_i) \in R$ exists as shown in Algorithm 5.5 which runs in $\omega(n)$ time.

Algorithm 5.4 *Reflexivity Test1*

```
1: procedure REFLEX_TEST1(R: Relation)
2:    for i = 1 to n do
3:       for j = 1 to n do
4:          if a_i = b_j then continue
5:             return(not_reflex)
6:          end if
7:       end for
8:    end for
9:    return(reflex)
10: end procedure
```

This algorithm runs in $O(n^2)$ time. We have not stated how to represent the relation R in a structure to be accepted by a computer in this pseudocode. Let us attempt to write the code in a closer structure to a computer language using the binary matrix representation of a relation. In this case, all we need to do is to check for all 1s in the diagonal of the matrix as shown in Algorithm 5.5 which runs in $O(n)$ time.

5.1.4.2   Symmetric and Antisymmetric Relations
A symmetric relation R is the one in which for any pair (a, b) that is contained in R, (b, a) is also a member of R.

Definition 5.4 (*symmetric, antisymmetric*) A relation R on a set A is symmetric if and only if $\forall a, b \in A$, $(b, a) \in R$ when $(a, b) \in R$. In a symmetric relation, aRb is commonly written as $a \sim b$, $a \approx b$, or $a \equiv b$.

Algorithm 5.5 *Reflexivity Test2*

1: **procedure** REFLEX_TEST2(M: Relation matrix)
2:     **for** $i = 1$ to n **do**
3:         **if** $M[i, i] \neq 1$ **then**
4:             **return** (*not_reflex*)
5:         **end if**
6:     **end for**
7:     **return** (*reflex*)
8: **end procedure**

$$\forall a \forall b : (a, b) \in R \rightarrow (b, a) \in R$$

A relation R is *antisymmetric* if and only if there is not a single pair (a, b) such that $(a, b) \in R$ and $(b, a) \in R$.

$$\forall a \forall b : ((a, b) \in R \wedge (a \neq b)) \rightarrow (b, a) \notin R$$

Remark 5.1 A relation R may be both symmetric and antisymmetric, or may not be symmetric and not antisymmetric at the same time.

Example 5.1.7 Let a relation on set $A = \{a, b, c, d\}$ be $R = \{(a, b), (b, c), (b, a),$ $(b, d), (d, b), (c, d), (d, c)\}$. R is not symmetric because $(b, c) \in R$ and $(c, b) \notin R$. Finding at least one such pair $((b, c)$ in this example) is enough to deduce R is not symmetric, however, R is not antisymmetric either because of the existence of symmetric pairs.

An algorithm to test symmetric property of a relation is depicted in Algorithm 5.6.

Algorithm 5.6 *Symmetry Test1*

1: **procedure** SYMMETRY_TEST1(R: Relation)
2:     **for** $i = 1$ to n **do**
3:         **for** $j = 1$ to n **do**
4:             **if** $(a_i, b_i) \in R$ and $(b_i, a_i) \in R$ **then** continue
5:                 **return** *not_symmetric*
6:             **end if**
7:         **end for**
8:     **end for**
9:     **return** *symmetric*
10: **end procedure**

We can have the matrix version of this algorithm as depicted in Algorithm 5.7 where we check whether $(b, a) \in R$ whenever $(a, b) \in R$. A 1 in matrix entry for

(a, b) means we need to check whether (b, a) entry in the matrix is also 1. This algorithm requires $O(n^2)$ operations using two nested *for* loops.

Algorithm 5.7 *Symmetry Test2*

1: **procedure** SYMMETRY_TEST2(M: Relation matrix)
2:     **for** $i = 1$ to n **do**
3:         **for** $j = 1$ to n **do**
4:             **if** $M[i, j] = 1$ **then**
5:                 **if** $M[j, i] \neq 1$ **then**
6:                     **return** *not_symmetric*
7:                 **end if**
8:             **end if**
9:         **end for**
10:     **end for**
11:     **return** *symmetric*
12: **end procedure**

5.1.4.3 Transitive Relation

A transitive relation relates three elements of a relation as follows.

Definition 5.5 (*transitive*) A relation R on a set A is transitive if and only if $\forall a, b, c \in A$: if $(a, b) \in R$ and $(b, c) \in R$ then $(a, c) \in R$.

$$\forall a \forall b \forall c : ((a, b) \in R \wedge (b, c) \in R) \rightarrow (a, c) \in R$$

Example 5.1.8 Let a relation on set $A = \{a, b, c, d\}$ be $R = \{(a, b), (b, c), (a, d), (b, d), (d, b), (c, d), (d, c)\}$. R is not transitive since $(a, b) \in R$ and $(b, c) \in R$ but $(a, c) \notin R$. Finding at least one such triplet (a, b, c) is enough to deduce R is not transitive. A closer look at R shows $(a, c) \notin R$ although $(a, d), (d, c) \in R$, thus (a, c, d) is another triplet that disproves transitivity.

5.1.4.4 Determining Relation Property from the Digraph and Matrix

It is possible to detect a property of a relation by looking at its graph or its binary matrix representation as follows.

- *Reflexive*: If each node of the graph contains a self loop, then R is reflexive since $\forall a \in A, (a, a) \in R$. Thus, the graph of relation R in Fig. 5.2 is reflexive. In the binary matrix (M) representation, we would have all 1's in the main (principal or leading) diagonal of the matrix M.
- *Symmetric*: In order to have a symmetric relation R, there should be a directed edge in both directions (from a to b and b to a) for each vertex pair (a, b) in the graph.

The binary matrix M should be symmetric with respect to its leading diagonal to have R symmetric. In other words, $M[i, j] = M[j, i]$ for each row i and column j of M.

- *Antisymmetric*: An antisymmetric relation manifests itself by the lack of any bidirectional edges in the relation graph. Having such an edge implies the existence of $(b, a) \in R$ when $(a, b) \in R$ for at least one such pair which means R is not antisymmetric. In matrix notation, if there exists one entry in M such that $M[i, j] = M[j, i]$, then R is not antisymmetric.
- *Transitive*: A transitive relation in the digraph of a relation has edges such that whenever there is a directed edge from any node a to a node b, and b to another node c in the graph, there is also an edge from a to c.

5.1.5 Equivalence Relations and Partitions

Equivalence relations provide grouping of similar elements of a set. These elements behave similarly and form *partitions* of the set.

Definition 5.6 (*equivalence*) A relation R on a set A is an equivalence relation if R is reflexive, symmetric and transitive.

Definition 5.7 (*equivalence class*) Let R be a relation on a set A and let $a \in A$. The equivalence class shown by $[a]$ is the set of all elements of A that satisfy relation R. In other words, any $b \in A$ that is related to a by R is in the same equivalence class as a. This relation can be stated as follows:

$$[a] = \{b \in A | bRa\}$$

Example 5.1.9 Let relation R be defined on people as $a R b$ if and only if a person a lives in the same city as a person b. We will assume a person lives only in one city and check whether this is an equivalence relation.

- A person a lives in the same city as a, hence R is reflexive.
- If a person a lives in the same city as a person b, then b lives in the same city as a, meaning R is symmetric.
- If a person a lives in the same city as a person b and b lives in the same city as a person c, then a lives in the same city as c, thus R is transitive.

We can therefore state that R is an equivalence relation that divides the people to equivalence classes, each class being the people living in the same city.

Example 5.1.10 Let relation R be defined on set $A = \{1, 2, 3, 4\}$ as follows:

$$R = \{(1, 1), (1, 3), (2, 2), (2, 4), (3, 1), (3, 3), (4, 2), (4, 4)\}$$

Find whether R is an equivalence relation.

Solution: Let us check all properties of an equivalence relation on R.

- $\forall a \in A$, $(a, a) \in R$; therefore, R is reflexive. Note that we needed $(1, 1)$, $(2, 2)$, $(3, 3)$ and $(4, 4)$ to be contained in R for R to be reflexive.
- Checking manually shows that $\forall a, b \in A$; if $(a, b) \in R$ then $(b, a) \in R$, thus R is symmetric.
- Again, checking shows $\forall a, b \in A$; if $(a, b) \in R$ and $(b, c) \in R$, then $(a, c) \in R$, thus R is transitive.

Since R satisfies all of the equivalence properties, we can state that R is an equivalence relation.

Definition 5.8 (*partition*) Let R be an equivalence relation on A and $a, b \in A$. Then $[x] = [y]$ if and only if $x R y$.

Example 5.1.11 A relation R is defined on $\mathbb{Z}$ by $a R b$ if $a + b$ is even. Show that R is an equivalence relation.

Solution: We need to check reflexive, symmetric and transitive properties for R to be an equivalence relation.

- *Reflexivity*: Let $a \in \mathbb{Z}$, then $a + a = 2a$ is an even integer. Therefore $a R a$ and R is reflexive.
- *Symmetry*: When $a R b$, we need to show $b R a$ holds. If $a + b = 2k$, then $b + a = 2k$. *Transitivity*: Assume $a R b$ and $b R c$, then $a R c$ should be valid.

Let $a + b = 2k$, $b + c = 2m$ for integers k and m. Then $a + c = 2(k + m - 1)$ is even.

5.1.6 Order

A *tuple* is a set of objects. When the length of a tuple is 1, it is called a *singleton*, length 2 is a *pair*, 3 is a *triple* and a tuple with length n is called an n-tuple. Different than sets, tuples can contain an object more than once and the objects in a tuple have a certain order. Commonly, a tuple is specified as (a, b, c) or $< a, b, c >$. We can specify *partial order* or *total order* relation. A partial order is a type of relation that orders the elements of a set.

Definition 5.9 (*partial order*) A relation R that is reflexive, antisymmetric and transitive is called a *partial order*. A set having a partial is called a *partially ordered set* or a *poset*.

Fig. 5.4 Composite relation

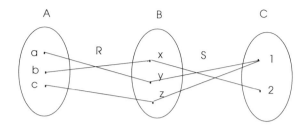

Example 5.1.12 $\forall a, b \in N$, let relation R be defined as $(a, b) \in R$ such that a divides b. Show that R is partial order.

Solution: Let us check all properties of an equivalence relation on R.

- $\forall a \in N, (a, a) \in R$ a divides a; therefore, R is reflexive.
- $\forall a, b \in N$, if $(a, b) \in R$, $(b, a) \notin R$ since if a divides b, then b does not divide a. Hence, R is antisymmetric.
- $\forall a, b \in N$, if a divides b and b divides c, then a divides c. Thus, R is transitive.

We can therefore conclude that R is a partial order.   □

Definition 5.10 (*total order*) A total order on a set A is a partial order relation R on A in which for every pair $(a, b) \in A$, either $a \leq b$ or $b \leq a$. For example, the set of real numbers ordered by $\leq$ is totally ordered.

5.1.7 Composite Relation

Definition 5.11 (*composite relation*) Let R be a relation from a set A to set B, and S be a relation from set B to a set C. The composition of R and S, shown by $S \circ R$ is the relation from the set A to set C as follows:

$$S \circ R = \{(a, c) | (a, b) \in R \text{ and } (b, c) \in S \text{ for some } b \in B\}$$

In other words, $a(S \circ R)b$ if and only if there exists some element $b \in B$ such that aRb and bSc. Graphically, a composite relation $R \circ S$ can be drawn as in Fig. 5.4 where $a(S \circ R)1$, $b(S \circ R)2$, and $c(S \circ R)1$.

Example 5.1.13 Let relations R and S be defined as below. Find $S \circ R$.

$$R = \{(1, 1), (1, 3), (2, 2), (2, 4), (3, 1)\}$$

$$S = \{(1, 1), (3, 2), (2, 1), (4, 3)\}$$

Solution: Applying the composition relation property, the following can be stated.

- $(1, 1) \in R$ and $(1, 1) \in S \to (1, 1) \in S \circ R$.
- $(1, 3) \in R$ and $(3, 2) \in S \to (1, 2) \in S \circ R$.
- $(2, 2) \in R$ and $(2, 1) \in S \to (2, 1) \in S \circ R$.
- $(2, 4) \in R$ and $(4, 3) \in S \to (2, 3) \in S \circ R$.
- $(3, 1) \in R$ and $(1, 1) \in S \to (3, 1) \in S \circ R$.

5.1.8 n-Ary Relations

An n-ary relation is a subset of a Cartesian product of n sets.

Definition 5.12 (*n-ary relation*) Given sets A_1, A_2, ..., A_n, an n-ary relation R on $A_1 \times A_2 \times ... \times A_n$ is a subset of $A_1 \times A_2 \times ... \times A_n$. If $A_1 = A_2 = ... = A_n$, the n-ary relation is defined on A, that is on subsets of $A \times A \times ... A$ which is A multiplied n times by itself. The sets A_i are called the *domains* of R and the degree of R is n.

Example 5.1.14 Let R be a 3-ary relation on $\mathbb{Z} \times \mathbb{Z} \times \mathbb{Z}$ consisting of triplets (a, b, c) such that $a < b < c$. For example, $(1, 2, 6) \in R$, $(2, 8, 12) \in R$ but $(5, 3, 9) \notin R$

The n-ary relations are the basis of relational database management systems as we will see.

5.1.9 Transitive Closure

A relation R may not be transitive due to the lack of some ordered pairs. Adding the least number of ordered pairs to the relation R results in a relation R^t which is called the *transitive closure* of R.

Definition 5.13 (*transitive closure*) Let R be a relation on set A. The set R^t is called the transitive closure of R if the following is valid:

- R^t is transitive
- $R \subseteq R^t$
- If S is another transitive relation of R, then $R^t \subseteq S$.

Example 5.1.15 Let $A = \{1, 2, 3, 4\}$ and the relation R on A to be as below:

$$R = \{(1, 2), (2, 3), (2, 4), (4, 1)\}$$

Find the transitive closure of R^t of R.
Solution: We find $(1, 2)$ and $(2, 3) \in R$ but $(1, 3) \notin R$. Similarly $(2, 1) \notin R$ and $(4, 2) \notin R$. Therefore, R^t can be formed to consist at least of the following elements:

$$R = \{(1, 1), (1, 2), (1, 3), (1, 4), (2, 1), (2, 2), (2, 3), (2, 4), (4, 1), (4, 3), (4, 4)\}$$

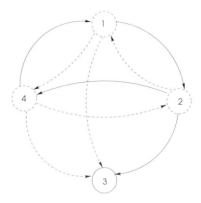

Fig. 5.5 Transitive closure of a relation with added pairs shown by dashed lines

Table 5.1 A database table

Number	Name	Dept.	GPA
1234	John	EE	3.12
5678	Sean	CS	2.81
9123	Nuri	Math	3.45
4567	Bob	CS	3.20

Alternatively, we can form the directed graph of Fig. 5.5 of relation R and check whether there is an arc from a node a to c when there is an arc from a to b, and b to c. The dotted lines are the added pairs to relation R to form the transitive closure R^t of R.

5.1.10 Database Applications

A *database* of a computer system consists of organized information stored in a computer. For example, a student database at a university typically will have numbers, age, name, courses she has taken and grades of a student. A database management system (DBMS) contains software modules to search, modify and interpret information in a database. A DBMS is basically a collection of tables each of which is organized by *columns* or *fields*. The rows in a DBMS are commonly called *attributes* of *records*, with each record consisting of a row in the table. A record is basically an n-tuple in a table with n columns. Table 5.1 depicts a simplified database of students in a university.

A database based on n-ary relation is called a *relational database management system* (RDBMS). Informally, data in tables can be accessed and reorganized without changing the structure of the original table in such a database. Each field in a RDBMS is a domain of the n-ary relation and a domain that uniquely identifies a record is

Table 5.2 Table 5.1 after projection

Name	GPA
John	3.12
Sean	2.81
Nuri	3.45
Bob	3.20

called a *primary key*. For example, the *Number* field in Table 5.1 is the primary key for this database. The following operations are common in a RDBMS.

- **Unary Operations**: These take one table as input and produce another one

  - *Selection*: The selection operator is used to filter rows based on some conditions. For example, let condition $C1$ be "Dept=Math" and $C2$ be "GPA>3.0". Then, $s_{C1 \wedge C2}$ will result in (9123, Nuri, Math, 3.45) for the example database in Table 5.1 filtering all other rows.
  - *Projection*: The projection operator $P_{i_1,\dots,i_m}$ defined on n-ary relation R as $P_{i_1,\dots,i_m} = \{(a_1, .., a_n) : \exists (b_1, .., b_k) \in R\}$ such that $a_1 = b_{i_1}, \dots, a_m = b_{i_m})\}$. For example, $P_{2,4}$ applied to Table 5.1 will result in Table 5.2.

- **Binary Operations**: We have two input tables this time and a resulting table is produced at the end of the operation.

  - *Join Operation*: Two n-ary relations R and S are input and an n-ary relation $R \cup S = \{(a_1, .., a_n) : (a_1, .., a_n) \in R \vee (a_1, .., a_n) \in S\}$ is produced.
  - *Difference Operation*: Two n-ary relations R and S are input and an n-ary relation $R - S = \{(a_1, .., a_n) : (a_1, .., a_n) \in R \wedge (a_1, .., a_n) \notin S\}$ is produced.
  - *Cartesian Product*: An n-ary relation R and an m-ary relation S are input and a relation $R \times S = \{(a_1, .., a_n, b_1, ..., b_m) : (a_1, .., a_n) \in R \wedge (b_1, .., a_m) \in S\}$ is produced.

5.2 Functions

A function is a special type of relation which basically maps one or more elements of a set to an element of another set. Commonly, a function $f(x)$ is an expression of a variable labeled x such as,

$$x^2 + 3x + 5, \quad \log x, \quad 2^x$$

Fig. 5.6 An example
function

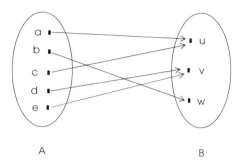

$$A \qquad\qquad\qquad\qquad\qquad B$$

Definition 5.14 (*function*) Given two sets A and B, a function f from A to B denoted by $f : A \rightarrow B$, is a relation over $A \times B$ such that for any $(a, b) \in f$, there exists one and only one element $b \in B$.

The set A is called the *domain* and the set B is called the *codomain* of function f. A function is commonly defined in terms of its domain and its codomain as below.

$$f = \{(x, y) \in \mathbb{Z} \times \mathbb{Z} : y = x^2 - 3x + 5\}$$

The function $f(x)$ is commonly called "f of x"; "the value of f at x", "the output of f for the input x", or "the image of x under f". A function is sometimes called a *mapping* (from its domain to its codomain).

Example 5.2.1 Let $A = \{a, b, c, d, e\}$ and $B = \{u, v, w\}$ two sets which are related by the function $f(a) = f(c) = u$, $f(b) = w$ and $f(d) = f(e) = v$. the graphical representation of this function is depicted in Fig. 5.6. It can be seen that there are directed edges from every element of A to some element of B; an element of A has only one directed edge connected to it in line with the definition of a function.

A function can be *partial* or *total*. A partial function is defined for a subset of the domain and a total function is defined for all values of the domain. For example $f(x) = 1/x$, $x \neq 0$ defined on $\mathbb{R}$ is a partial function defined for all values on $\mathbb{R}$ except the 0 value. The function $f(x) = x^3$ however is a total function on $\mathbb{R}$ defined for all values in this set.

5.2.1 Composite Functions

Definition 5.15 (*composite function*) Let $f : A \rightarrow B$ and $g : B \rightarrow C$ be two functions. The *composite function* $g \circ f : A \rightarrow C$ is defined as follows:

$$g \circ f = \{(a, c) \in A \times C : (a, b) \in f \text{ and } (b, c) \in g \text{ for some } b \in B\}$$

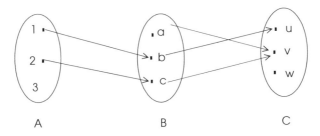

Fig. 5.7 Graph of $g \circ f$

Example 5.2.2 Let $f(x) = 2x + 3$ and $g(x) = x^2 + 1$ be defined on $\mathbb{R}$. Then,

$$g \circ f = g(f(x))$$
$$g(2x + 3) = (2x + 3)^2 + 1$$
$$= 4x^2 + 12x + 10$$

The composite function is not commutative, that is, $g \circ f \neq f \circ g$ in general. For the above example,

$$f \circ g = f(g(x))$$
$$g(2x + 3) = 2(x^2 + 1) + 3$$
$$= 2x^2 + 5$$

However, the composite function is associative; $h \circ (g \circ f) = (h \circ g) \circ f$. A composite function $(g \circ f)$ is depicted in Fig. 5.7.

5.2.2 Injection, Surjection and Bijection

Definition 5.16 (*injection, one-to-one*) A function $f : A \rightarrow B$ is called *injective* or *one-to-one* if for all $a, b \in A$, if $a \neq b$, then $f(a) \neq f(b)$.

In other words, if $f(a) = f(b)$, then $a = b$. This means there is exactly one directed edge between a pair of elements of the digraph representing the function. The sample function graph of Fig. 5.6 displays a non one-to-one function as there are two edges from set A to nodes u and v of set B. Note that some elements of the codomain may not be covered by a one-to-one function but any covered element of codomain should be the output of exactly one element of the domain. An injective function is depicted in Fig. 5.8a.

An algorithm to test whether a function is injective or not can be formed with the following logic. The binary matrix M of a function f should not have more than one 1 in any of its columns to be injective since otherwise two elements of the domain is mapped to the same element of the codomain of the function. The following matrix

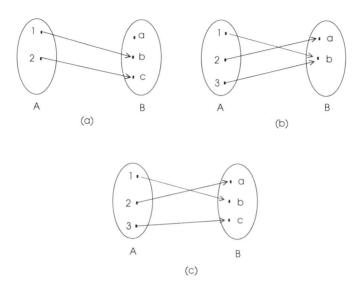

Fig. 5.8 **a** An injective function, **b** a surjective function, **c** a bijection

denotes an injective function $f = \{((1, b), (2, a), (3, c)\}$ from the set $A = \{1, 2, 3\}$ to set $B = \{a, b, c\}$ since each column has at most one 1.

$$
\begin{array}{c}
\begin{array}{ccc} a & b & c \end{array} \\
\begin{array}{c} 1 \\ 2 \\ 3 \end{array}
\begin{pmatrix} 0 & 1 & 0 \\ 1 & 0 & 0 \\ 0 & 0 & 1 \end{pmatrix}
\end{array}
$$

The algorithm to test injection shown in Algorithm 5.8 checks each row and whenever it encounters a 1 in column j, it checks to see whether there is another 1 in the same column. This algorithm requires $O(n^3)$ operations due to 3 nested *for* loops.

Example 5.2.3 Let $f(x) = 3x + 4$ be defined on $\mathbb{R}$. Find whether f is one-to-one. *Solution*: Let us select $a, b \in \mathbb{R}$ and test whether $f(a) = f(b)$ if and only if when $a = b$ (definition of one-to-one). Substitution yields $3a + 4 = 3b + 4$ and simplification gives $a = b$, thus f is one-to-one.   □

Definition 5.17 (*surjection, onto*) A function $f : A \rightarrow B$ is called *surjective* or *onto* if for all $b \in B$ there is t least one $a \in A$ with $f(a) = b$.

Example 5.2.4 Let $A = \{a, b, c, d, e\}$ and $B = \{u, v, w\}$ two sets which are related by the function $f(a) = f(c) = u$, $f(b) = w$ and $f(d) = f(e) = v$. Since all elements of B are covered, f is an *onto* function.

Algorithm 5.8 *Injection Test*

```
1: procedure INJECTION_TEST(M: function matrix)
2:    for i = 1 to n do
3:       for j = 1 to m do
4:          if M[i, j] = 1 then
5:             for k = j + 1 to n do
6:                if M[j, k] = 1 then
7:                   return(not_injective)
8:                end if
9:             end for
10:         end if
11:      end for
12:   end for
13:   return(injective)
14: end procedure
```

The matrix M representing a function f should have exactly one 1 in each of its row by the definition of function and the function being onto means all elements of the domain should be mapped. We can form an algorithm to test whether a function represented by M is onto or not by considering that each row should have one 1, as shown in Algorithm 5.9. We will assume there will not be more than one 1 in each row as the input is a legal function.

Algorithm 5.9 *Surjection Test*

```
1: procedure SURJECTION_TEST(M: function matrix)
2:    for i = 1 to n do
3:       for j = 1 to m do
4:          if M[i, j] = 1 then continue
5:             return(not_onto)
6:          end if
7:       end for
8:    end for
9:    return(onto)
10: end procedure
```

Definition 5.18 (*bijection*) A function $f : A \rightarrow B$ is called *bijective* if it is both injective (one-to-one) and surjective (onto).

Examples of these functions are depicted in Fig. 5.8 where the function in (a) is one-to-one as there is a mapping of each element of A in B; (b) is an onto function since every element of B is covered by the function and (c) is both one-to-one and onto.

5.2.3 Inverse of a Function

The inverse of a function can be defined similarly to the inverse of a relation.

Definition 5.19 (*function inverse*) The inverse of a function $f : A \to B$ is denoted by $f^{-1} = B \to A$ such that if $f(a) = b$, then $(b, a) \in f^{-1}$.

Based on the definition of a function, a total function has an inverse if and only if it is bijective. Let the identity function on a set A be $I_A(a) = a$, for all $a \in A$. Then, $f^{-1} \circ f = I_A$, and $f \circ f^{-1} = I_B$.

Example 5.2.5 Let $A = \{1, 2, 3\}$ and $B = \{a, b, c\}$ be two sets which are related by the function $f(1) = b$, $f(2) = a$ and $f(3) = c$ as in Fig. 5.8c. This function is bijective and let us check whether the above equalities hold. Visually, $f^{-1}(b) = 1$, hence $f^{-1} \circ f(1) = 1$ and this is valid for the other two members of set A. On the other hand, $f^{-1}(a) = 2$ and $f(2) = a$ and checking for b and c in B, we can deduce $f \circ f^{-1} = I_B$.

Example 5.2.6 Let $f : \mathbb{N} \to \mathbb{N}$ be defined as $f(x) = 2x + 1$. Let $f^{-1}(y) = x$, a practical way to find f^{-1} is to substitute y for x, and x for y in f and solve for y as below.

$$x = 2y + 1$$
$$y = \frac{x - 1}{2}$$

Let us check for an arbitrary number, say 2, $f(2) = 5$ and $f^{-1}(5) = (5 - 1)/2 = 2$.

5.2.4 Some Special Functions

Some functions are very common as they have a wide spectrum of applications. A sample of these well-known functions are the floor and ceiling functions, the factorial function reviewed previously, the exponential function and the logarithmic function.

Floor and Ceiling Functions
The floor and ceiling functions of a real variable are used to generate integer values from those variables.

Definition 5.20 (*floor function, ceiling function*) The *floor function* of a real number is the largest integer that is less than or equal to that number. This function of a variable x is shown as $\lfloor x \rfloor$. The *ceiling function* of a real number is the smallest integer that is greater than or equal to that number which is represented by $\lceil x \rceil$.

For example, $\lfloor 3.12 \rfloor = 3$, $\lceil 3.12 \rceil = 4$, $\lfloor 3 \rfloor = \lceil 3 \rceil = 3$, $\lfloor -0.25 \rfloor = -1$, $\lceil -0.25 \rceil = 0$.

Factorial Function

Factorial of a positive integer n is defined as the product of all integers from 1 to n.

Definition 5.21 (*factorial function*) The factorial of $n \in Z$ is denoted by $n!$ and,

$$n! = 1 \cdot 2 \cdot 3 \cdots n$$

For example, the factorial of $5 = 1 \cdot 2 \cdot 3 \cdot 4 \cdot 5 = 120$. We can form a procedure that input the value of n and returns its factorial as in Algorithm 5.10.

Algorithm 5.10 *Factorial Function*

1: **procedure** FACTORIAL(n: integer)
2:    $fact \leftarrow 1$
3:    **for** $i = 1$ to n **do**
4:        $fact \leftarrow fact \cdot i$
5:    **end for**
6:    **return** $fact$
7: **end procedure**

Exponential Function

Definition 5.22 (*exponential function*) The function $f(a, x) = a^x$ is defined as

$$a^m = a \cdot a \cdots a \text{ for } m \text{ times}$$

where $m \in Z$.

For example, $2^3 = 2.2.2 = 8$, $4^{-2} = 1/(4^2) = 1/16$. An exponent can be a rational number such as m/n. Then, $a^{m/n} = \sqrt[n]{a^m}$. A simple procedure that inputs the value of base a and the exponential m and returns a^m is shown in Algorithm 5.11.

Algorithm 5.11 *Exponential Function*

1: **procedure** EXPONENTIAL(a: base, m: exponential)
2:    $exp \leftarrow 1$
3:    **for** $i = 1$ to m **do**
4:        $exp \leftarrow exp \cdot a$
5:    **end for**
6:    **return** exp
7: **end procedure**

Logarithm

Definition 5.23 (*logarithm function*) Given $x > 0$ and $a > 0$ and $a \neq 1$, the logarithmic function $f(x)$ is defined as,

$$f(x) = y = \log_a x \text{ if and only if } x = a^y$$

where a is the base of the logarithm.

For example, $\log_{10} 100 = 2$ since $10^2 = 100$. Common logarithm bases are 100 and 2. Some important properties of a logarithmic function are as follows.

$$\log_a xy = \log_a x + \log_a y \text{ for } a, b \in \mathbb{R}, x > 0 \text{ and } y > 0 \qquad (5.1)$$

$$\log_a x^y = y \log_a x \text{ for } a, b \in \mathbb{R}, x > 0 \text{ and } y > 0. \qquad (5.2)$$

For example, $\log_{10} 10 \cdot 100 = \log_{10} 1000 = 3$ by the definition of the logarithmic function since $10^3 = 1000$. Based on Eq. 5.1, we can write $\log_{10} 10 + \log_{10} 100 = 1 + 2 = 3$ to give the same result. Consider another example, $\log_{10} 10^2 = \log_{10} 100 = 2$ and using Eq. 5.2, $\log_{10} 10^2 = 2 \log_{10} 10 = 2$ yields the same result.

5.3   Review Questions

1. What is meant by the union of two relations R_1 and R_2?
2. What is the difference between a unary relation and a binary relation?
3. What is meant by the inverse of a relation?
4. Compare symmetric and antisymmetric properties of a relation.
5. How can the reflexive property of a relation be determined from its graph representation?
6. How can the symmetric property of a relation be determined from its matrix representation?
7. What is an equivalence relation? Give an example of such a relation.
8. What is a partition?
9. What is an n-ary relation and how is this type of relation used in a relational database?
10. What is the main difference between a relation and a function?
11. What is a composite function?
12. Describe injection, surjection and bijection as referred to functions.

5.4 Chapter Notes

A binary relation from a set A to a set B associates elements of A to the elements of B. We initially reviewed basic relation types and properties in this chapter. A unary relation defined on a single set A consists of a subset of the cartesian product $A \times A$ whereas a binary relation is a subset of the cartesian product of two sets. Similarly, n-ary relation may be defined to be a subset of the cartesian product of n sets. A relation may be represented by a graph or a matrix. Union and intersection of two relations are defined similar to set structures. A relation may have reflexive, symmetric, antisymmetric and transitive properties. A relation that is reflexive, symmetric and transitive is called an equivalence class and a relation that is reflexive, antisymmetric and transitive is called a partial order. An n-ary relation forms the basis of a relational database management system.

A function f is basically a relation from a set A to another set B with the property that for any $(a, b) \in f$, there exists one and only one element $b \in B$. A function can be composite, consisting of two or more functions. A function may have injection, surjection and bijection properties. Some special functions include floor function, ceiling function, factorial function, exponential function and logarithm.

Exercises

1. Let $A = \{1, 3, 6, 10\}$, $B = \{2, 3, 5, 6\}$ and R be $a \equiv b \pmod{m}$ with $a \in A$ and $b \in B$. Find the elements of R and R^{-1} and draw the graphical representations of R and R^{-1}.
2. Let $A = \{1, 2, 3, ..., 5\}$ and 3-ary relation R by (a, b, c) if and only if $a < b$ and $b < c$. List all of the elements of R.
3. Let $A = \{1, 3, 5, 7, 9, 10, 13, 17\}$, $B = \{2, 3, 5, 6\}$ and R be defined as aRb if a is prime and $b = 2a$. Express R as ordered pairs.
4. A relation R is defined on $\mathbb{Z}$ by aRb if $a + 3b$ is even. Prove that R is an equivalence relation and find the equivalence classes of R.
5. A relation R on a set A is given as follows. For any $a, b \in A$, aRb if $f(a) = f(b)$ for any function $A \to B$. Prove that R is an equivalence relation.
6. Let A be the power set of a finite set A and R be a relation such that for $S_1, S_2 \in \mathcal{P}(A)$, $S_1 R S_2$ if $|S_1| = |S_2|$. Show that R is an equivalence relation on $\mathcal{P}(A)$.
7. Write the pseudocode of an algorithm procedure to test transitivity of a relation and show its running on relation $R = \{(1, 2), (2, 3), (1, 3), (2, 2), (3, 4), (1, 4)\}$.
8. Write the pseudocode of an algorithm to form the transitive closure of a relation. Show the working of this algorithm for the relation $R = \{(1, 2), (2, 3), (2, 4), (4, 1), (2, 1)\}$.
9. Write the pseudocode of an algorithm procedure to test whether a given relation R with its binary matrix is a legal function or not.
10. Write the pseudocode of an algorithm to test whether a given function f is bijective or not.
11. Prove Eqs. 5.1 and 5.2.

Sequences, Induction and Recursion

6

We review three related topics in this chapter: sequences, induction and recursion. A sequence is an ordered list of elements and is a function from natural number set to a set of real numbers. Summation of sequences may be defined in various ways. Arithmetic sequences have terms with a common difference from the preceding one and geometric series have terms with a common ratio. Induction is a powerful proof method which has a wide range of applications. Recursion is the process of defining an object in terms of itself. We can have recursive relations, recursive sets, recursive functions and recursive algorithms. We review sequences, the basic induction method and recursion with a brief introduction to recursive algorithms in this chapter.

6.1 Sequences

A sequence is basically an ordered list of real numbers with an initial term. A *finite sequence* has a limited number of terms with a final term whereas an *infinite sequence* has infinite number of elements and does not have a final element. We are mostly interested in finite sequences and their properties. A formal definition of a sequence is as follows.

Definition 6.1 (*sequence*) A *sequence* is a function from $\mathbb{N}$ into a set A of real numbers. A sequence is commonly shown as $a_1, a_2, .., a_n$ and we say that such a sequence is indexed by integers.

In other words, the domain of a sequence is the set of natural numbers and the output is a subset of the real numbers. A sequence may contain the same term more than once and unlike a set, a sequence is ordered.

© Springer Nature Switzerland AG 2021
K. Erciyes, *Discrete Mathematics and Graph Theory*, Undergraduate Topics in Computer Science, https://doi.org/10.1007/978-3-030-61115-6_6

Example 6.1.1 Find the first 5 elements of the sequence $a_n = (-1)^n$ where $n = 0, 1, 2, 3, ...$

Solution: $1, -1, 1, -1, 1$. This sequence is an example of an alternating sequence.
A *closed form* representation of a sequence is stated as $a_n = f(n)$, $\forall n \geq 1$ where $f(n)$ is some function of n such that $f : \mathbb{N} \to \mathbb{R}$.

Example 6.1.2 Find the closed form for the sequence $1, 2, 4, 8,...$.

Solution: We can see that the terms of the sequence are the powers of 2 as $2^0, 2^1, 2^2$, ... which can be written as 2^n. Note that n starts from 0 in this sequence.

Example 6.1.3 Find the closed form for the sequence $2, 5, 8, 11, 14, 17...$
Solution: This sequence can be specified in closed form as $a_n = 3n - 1$. For example, $a_4 = 3 \cdot 4 - 1 = 11$ and $a_6 = 3 \cdot 6 - 1 = 17$.

6.1.1 Summation

The Greek symbol sigma ($\sum$) is used to define the sum of a sequence: For example,

$$\sum_{i=m}^{n} a_i$$

is the sum of a sequence with elements $a_m + a_{m+1} + ... + a_n$. The variable i in the above equation is called the *index* of the summation, m is the lower limit and n is the upper limit of the index.

Example 6.1.4 Find the sum of the first 5 terms of the sequence $a_n = n^2$ where $n = 1, 2, 3...$. *Solution*: The formula for the sequence is,

$$\sum_{i=1}^{5} a_i = \sum_{i=1}^{5} i^2 = 1 + 4 + 9 + 16 + 25 = 55$$

$\square$

Example 6.1.5 Express the sum of the first 10 terms of the sequence $a_n = 2n - 1$ for $n = 1, 2, 3,..., 10$ and find the value of the summation for these terms.
Solution: This sequence is the list of positive odd numbers and can be written as,

$$\sum_{i=1}^{10} 2i - 1 = 1 + 3 + 5 + 7 + 9 + 11 + 13 + 15 + 17 + 19 = 100$$

$\square$

Shifting an index may provide a more readable and workable summation as shown below when j is substituted for $i - 1$.

$$\sum_{i=1}^{n}(i - 1) = \sum_{j=0}^{n-1} j$$

There are cases when the summation is defined over a particular set as in the example below. Let S be $\{1, 2, 3\}$, then the following summation finds the sum of elements of this set. Note that we could have written $i \in \{1, 2, 3\}$ without explicitly specifying the set S.

$$\sum_{i \in S} i = 1 + 2 + 3 = 6$$

Properties of Summation
Summations can be merged and split as follows.

- Two summations can be merged into one when the index variables and their ranges are the same, as follows.

$$\sum_{i=m}^{n} a_i + \sum_{i=m}^{n} b_i = \sum_{i=m}^{n}(a_i + b_i)$$

- A summation can be split into two summations by dividing the original range into two subranges as below. Note that the sum of the two ranges equals the original range.

$$\sum_{i=m}^{n} a_i = \sum_{i=m}^{k} a_i + \sum_{i=k+1}^{n} a_i$$

- A double summation is denoted by a double $\sum$ sign and interchange of sum symbols is possible when indices of sums are independent as shown below.

$$\sum_{i=1}^{n} \sum_{j=1}^{m} ij = \sum_{j=1}^{m} \sum_{i=1}^{n} ij$$

6.1.2 Arithmetic Sequence and Series

Definition 6.2 (*arithmetic sequence*) An arithmetic sequence is a list where each term a_n of this sequence has a *common difference* value d from the previous one such that $a_n = a_{n-1} + d$. An arithmetic sequence with n elements has the form,

$$a, a + d, a + 2d, ..., (a + (n - 1)d)$$

where a is the *first term* and d is the *common difference* of the sequence.

For example; $-2, 1, 4, 7, 10 ..$ is an arithmetic sequence with $a = -2$ and $d = 3$. We can see that the nth element of an arithmetic series with first element a_1 is,

$$\forall n \geq 0, \ a_n = a_1 + (n-1)d$$

For the above example, $a_5 = -2 + (5-1)3 = 10$.

Definition 6.3 (*arithmetic series*) An *arithmetic series* is obtained by summing the elements of an arithmetic sequence.

Theorem 4 *The sum of the first n terms of an arithmetic series with the first term a_1 and the common difference d is,*

$$S_n = \frac{n(a_1 + a_n)}{2}$$

Proof Let us write the sum S_n of the first n terms of this series from left to right and right to left and add both sides as below.

$$S_n = a + (a+d) + (a+2d) + \ldots + (a + (n-1)d)$$
$$S_n = (a + (n-1)d) + (a + (n-2)d) + \ldots + a$$
$$2S_n = n(2a + (n-1)d)$$

Thus,

$$S_n = \frac{n(2a + (n-1)d)}{2} \tag{6.1}$$

substituting a_n for $a + (n-1)d$ in this formula yields,

$$S_n = \frac{n(a + a_n)}{2} \tag{6.2}$$

Example 6.1.6 Find the sum of the first 5 terms of the arithmetic sequence $-2, 3, 8, 13, 18...$.
Solution: The common difference d is 5 and the first term a is -2 in this sequence. Applying the sum formula of Eq. 6.1 yields,

$$S_5 = \frac{5((2 \cdot -2) + (4 \cdot 5))}{2} = 40$$

□

Example 6.1.7 Find the number of terms of the following series to yield the stated sum.

$$3 + 9 + 15 + \ldots = 75$$

Solution: The common difference d is 6 and the first term a is 3 in this sequence. Applying the sum formula of Eq. 6.1 and solving for n yields,

$$75 = \frac{n(2 \cdot 3 + 6(n-1))}{2}$$
$$150 = n(6 + 6n - 6)$$
$$= 6n^2$$
$$n^2 = 25$$

Since n can not be negative, $n = 5$. Note that closed form for this sequence is $6n - 3$ for $n = 1, 2, \ldots$. We could have used Eq. 6.2 to yield the same result in a simpler way as below. However, we need to work out the closed form in this case.

$$75 = n\frac{3 + (6n-3)}{2}$$
$$150 = 6n^2$$
$$n^2 = 25$$
$$n = 5$$

6.1.3 Geometric Sequence

Definition 6.4 (*geometric sequence*) Each successive term in a geometric sequence is obtained by multiplying the previous term with a constant r called the *common ratio*. This sequence can be written as,

$$a, ar, ar^2, \ldots, ar^n$$

For example; 1, 3, 9, 27... is a geometric sequence with $r = 3$. The nth element of a geometric sequence is,

$$\forall n \geq 0, \ a_n = ar^{n-1}$$

For this example, $a_4 = 1(3^{4-1}) = 27$.

Definition 6.5 (*geometric series*) A *geometric series* is obtained by summing the elements of a geometric sequence.

Theorem 5 *The sum S_n of the first n terms of an arithmetic series with the first term a and common factor r is,*

$$S_n = a\frac{(1 - r^n)}{1 - r} \text{ with } r \neq 1 \tag{6.3}$$

Proof Writing the sum S_n of the first n terms of this series and multiplying the sum with the common factor r provides the following,

$$S_n = a + (ar) + (ar^2) + \ldots + (ar^{n-1})$$
$$r\,S_n = (ar) + (ar^2) + \ldots + (ar^n)$$

Subtracting the first equation from the second one yields,

$$r\,S_n - S_n = ar^n - a = a(r^n - 1)$$
$$(r - 1)S_n = a(r^n - 1)$$

Thus,

$$S_n = a\frac{(r^n - 1)}{r - 1} = a\frac{(1 - r^n)}{1 - r} \text{ with } r \neq 1 \tag{6.4}$$

$\square$

Example 6.1.8 Find the sum of the first 4 terms of the geometric sequence 2, 6, 18, 54, ...

Solution: The common ratio r is 3 and the first term a is 2 in this sequence. Applying the sum formula of Eq. 6.4 yields,

$$S_4 = 2\frac{(1 - 3^4)}{1 - 3} = 80$$

$\square$

6.1.4 Product Notation

A series of values may be multiplied using a similar notation to sum, this time with the Π symbol as in the example below,

$$\prod_1^n i = 1 \cdot 2 \cdot 3 \cdots n$$

which specifies the product of integers from 1 to n. This expression is called the *factorial* of n and denoted as $n!$ as noted previously. Two products can be combined into one as follows.

$$\left(\prod_{i=n}^m a_i\right)\left(\prod_{i=n}^m b_i\right) = \prod_{i=n}^m (a_i \cdot b_i)$$

6.1.5 Big Operators

Similar to summation, various other operators may be defined over a sequence of statements as shown below. The big AND and big OR operators are commonly defined over propositions and the big union and the big intersection operators are defined over sets.

- **Big And**:

$$\bigwedge_{x \in S} P(x) \equiv P(x_1) \wedge P(x_2) \wedge P(x_3) \wedge \ldots \equiv \forall x \in S : P(x)$$

- **Big Or**:

$$\bigvee_{x \in S} P(x) \equiv P(x_1) \vee P(x_2) \vee P(x_3) \vee \ldots \equiv \exists x \in S : P(x)$$

- **Big Union**:

$$\bigcup_{i=1}^{n} A_i = A_1 \cup A_2 \cup A_3 \cup \ldots \cup A_n$$

- **Big Intersection**:

$$\bigcap_{i=1}^{n} A_i = A_1 \cap A_2 \cap A_3 \cap \ldots \cap A_n$$

6.2 Induction

Mathematical induction is a powerful proof method that can be used for very diverse applications, including proving the correctness of algorithms. Let us consider falling dominoes as an example to illustrate how induction works. We would consider the following when planning a domino experiment to have all dominoes in a chain fall.

1. The first domino will fall when we push it.
2. If a domino falls, then the next one in the chain will fall because it is placed close to its previous neighbor in chain.

We can now conclude that every domino in the chain of dominos will fall based on these two conditions. In order to formally state, Let $P(n)$ be a function defined over the set of natural numbers. The axiom of induction states that if $P(1)$ is true, and $P(n)$..., formally,

$$P(1) \wedge (\forall k \in N, (P(k) \rightarrow P(k+1)) \rightarrow \forall n P(n),$$

Based on the foregoing, induction consists of two steps.

- *Basis Step*: $P(1)$ is proven.
- *Inductive Step*: The conditional statement $P(k) \rightarrow P(k+1)$ is proven for all positive integers k.

Example 6.2.1 Prove that the sum of the first n positive integers is $n(n+1)/2$, that is, $\sum_{i=1}^{n} = n(n+1)/2$

Solution: The basis step is $\sum_{i=1}^{1} = 1(1+1)/2 = 1$ is true. Assume $P(k)$ is true for some integer k.

$$
\begin{aligned}
P(k+1) &= P(k) + (k+1) \\
&= \frac{k(k+1)}{2} + (k+1) \\
&= \frac{k(k+1) + 2k + 2}{2} \\
&= \frac{k^2 + 3k + 2}{2} \\
&= \frac{(k+1)(k+2)}{2}
\end{aligned}
$$

The final statement is exactly what we would get when $k+1$ is substituted in the general summation formula. $\square$

Example 6.2.2 Prove that sum of the first n odd positive integers, $\sum_{i=1}^{2n-1}$, is n^2 by induction.

Solution: The basis step is $\sum_{i=1}^{1} = 1^2 = 1$ is true. Let us write $P(k+1)$ as follows where kth odd integer is $2k-1$. Assume $P(k)$ is true for some integer k, then $P(k+1)$ is as follows.

$$
\begin{aligned}
P(k+1) &= P(k) + 2(k+1) - 1 \\
&= k^2 + 2(k+1) - 1 \\
&= k^2 + 2k + 1 \\
&= (k+1)^2
\end{aligned}
$$

Again, the final statement is exactly what we would get when $k+1$ is substituted in the formula $\sum_{i=1}^{2n-1} = n^2$. $\square$

Example 6.2.3 Prove that for every integer n, the sum of the squares of all integers from 1 to n is,
$$
\frac{n(n+1)(2n+1)}{6}
$$

Solution: The basis steps, $1^2 = (1 \cdot 2 \cdot 3)/6$ is true. We then assume,

$$1^2 + 2^2 + ... + k^2 = \frac{k(k+1)(2k+1)}{6}$$

and need to show,

$$1^2 + 2^2 + ... + (k+1)^2 = \frac{(k+1)(k+2)(2k+3)}{6}$$

where k is replaced by $(k+1)$.

$$
\begin{aligned}
1^2 + 2^2 + ... + (k+1)^2 &= (1^2 + 2^2 + ... + k^2) + (k+1)^2 \\
&= \frac{k(k+1)(2k+1)}{6} + (k+1)^2 \\
&= \frac{k(k+1)(2k+1)}{6} + \frac{6(k+1)^2}{6} \\
&= \frac{(k+1)(k(2k+1) + 6(k+1))}{6} \\
&= \frac{(k+1)(2k^2 + 7k + 6)}{6} \\
&= \frac{(k+1)(k+2)(2k+3)}{6}
\end{aligned}
$$

□

Example 6.2.4 Prove that for every integer n,

$$1 + 2 + 2^2 + 2^3 + ... + 2^n = 2^{n+1} - 1$$

Solution: The basis step is, $2^0 = 1 = 2^1 - 1$ is true. We assume the following

$$1 + 2 + 2^2 + 2^3 + ... + 2^k = \sum_{i=0}^{k} 2^i = 2^{k+1} - 1$$

need to show,

$$\sum_{i=0}^{k+1} 2^i = 2^{k+2} - 1$$

where k is replaced by $(k + 1)$.

$$\sum_{i=0}^{k} 2^i = 2^{k+1} - 1$$

$$\sum_{i=0}^{k} 2^i + 2^{k+1} = (2^{k+1} - 1) + 2^{k+1}$$

$$\sum_{i=0}^{k+1} 2^i = 2 \cdot 2^{k+1} - 1$$

$$= 2^{k+2} - 1$$

□

Example 6.2.5 Prove that for every positive integer n, $5n^2 - n + 4$ is even.
Solution:

- *Basis Step*: For $k = 1$, $5k^2 - k + 4 = 8$ is even.
- *Inductive Step*: Assuming $5k^2 - k + 4$ is even, we need to show $5(k+1)^2 - (k+1) + 4$ is even. By the induction hypothesis, let $5k^2 - k + 4 = 2m$ for some integer m.

$$5(k + 1)^2 - (k + 1) + 4 = 5(k^2 + 2k + 1) - (k + 1) + 4$$
$$= 5k^2 + 9k + 8$$
$$= (5k^2 - k + 4) + 10k + 4$$
by induction hypothesis
$$= 2m + 2(5k + 2)$$
let $5k + 2 = l$
$$= 2m + 2l = 2(m + l)$$

□

Remark 6.1 We could have used a similar calculation for the inductive case for $5n^2 - n + 7$ to deduce the result is even, however, $5n^2 - n + 7$ is always odd. The fallacy in this case is that we have not specified a basis step, in fact, we can not find a base case for this expression to be even.

6.2.1   Proving Inequalities

Induction may be used conveniently to prove inequalities. For example, to prove $\forall n \in \mathbb{N}$, $2^n > n$; we apply induction method as follows.

- *Basis Step*: Let $k = 1$, then $2^1 > 1$ is true.

- *Inductive Step*: Assume $P(k) : 2^k > k$ is true, we need to show $2^{k+1} > k + 1$ is true for $k \geq 1$.

$$
\begin{aligned}
2^{k+1} = 2 \cdot 2^k > 2k \qquad &\text{by the induction hypothesis} \\
> k + k \geq k + 1 \qquad &\text{since } k \geq 1 \\
> k + 1
\end{aligned}
$$

□

Example 6.2.6 Prove that for every integer $n \geq 5$, $2^n > n^2$.
Solution:

- *Basis Step*: Let $k = 5$, then $2^5 = 32 > 5^2 = 25$ is true.
- *Inductive Step*: Assume $P(k) : 2^k > k^2$ is true when $k \geq 5$, we need to show $2^{k+1} > (k + 1)^2$ is true for $k \geq 5$

$$
\begin{aligned}
2^{k+1} = 2 \cdot 2^k > 2k^2 \qquad &\text{by the induction hypothesis} \\
> k^2 + k^2 \geq k^2 + 5k \qquad &\text{since } k \geq 5 \\
> k^2 + 2k + 3k \geq k^2 + 2k + 1 \\
> k^2 + 2k + 1 = (k + 1)^2
\end{aligned}
$$

□

6.3 Strong Induction

Strong induction is similar to simple (weak) induction with one major difference; in order to prove $P(k + 1)$, we will assume $P(1), P(2), \dots, P(k)$ are all true. This method is also called the *strong principle of mathematical induction*. Formally, this method consists of two steps.

- *Basis Step*: $P(1)$ is proven.
- *Inductive Step*: $\forall k \in \mathbb{N}, (P(1) \wedge P(2) \wedge \dots \wedge P(k)) \rightarrow P(k + 1)$ is true.

Strong induction, sometimes also called the *second principle of induction*, is basically equivalent to weak (first principle of) induction but it may be simpler to use than the first one to prove certain propositions.

Theorem 6 (fundamental theorem of arithmetics) *Show that any integer $n \geq 1$ is either a prime number or can be written as the product of primes.*

For example, $18 = 2 \cdot 3 \cdot 3$; $24 = 2 \cdot 2 \cdot 2 \cdot 3$.

Proof We have the following:

- *Basis Step*: $P(2)$ can be written as the product of itself.
- *Inductive Step*: We have two cases as follows.

  - $k + 1$ is prime, then it is the product of itself.
  - $k + 1$ is a composite number and there exists two positive integers a and b, $2 \leq a \leq b \leq k$ such that,
  $$k + 1 = a \cdot b$$

  These two integers can be written as the product of primes by the induction hypothesis, therefore, $k+1$ is the product of primes. Note that we have considered both $P(a)$ and $P(b)$ to be true to prove $P(k + 1)$ as in the definition of strong induction.

  □

Fibonacci Numbers

Fibonacci numbers named after the Italian mathematician Fibonacci is the sequence;

$$0, 1, 1, 2, 3, 5, 8, 13, 21, 34, 55, 89, 144, 233, \dots.$$

as noted. If each number is represented by f_n, with $f_0 = 0$ and $f_1 = 1$,

$$f_n = f_{n-1} + f_{n-2}$$

Example 6.3.1 Let the nth Fibonacci number be f_n, prove that $f_n < 2^n$.

Solution: We can see that $f_1 = 1 < 2$ and $f_2 = 1 < 2^2 = 4$. Thus, the statement holds for $n = 1$ and $n = 2$. Now, using strong induction, assume $f_m < 2^m, \forall m \in \mathbb{N}^+$ when $m \leq k$. We can then state the following for $k \geq 2$,

$$\begin{aligned}
f_{k+1} &= f_k + f_{k-1} & \text{definition of Fibonacci numbers} \\
&< 2^k + 2^{k-1} & \text{by the inductive hypothesis} \\
&< 2^k + 2^k & \text{since } 2^{k-1} < 2^k \text{ for } k \geq 1 \\
&< 2^{k+1}
\end{aligned}$$

Therefore, $f_n < 2^n$ for al positive integers n. Note that we assumed both $f_k < 2^k$ and $f_{k-1} < 2^{k-1}$ to prove $f^{k+1} < 2^{k+1}$.   □

6.4 Recursion

Defining an object in terms of itself may provide a convenient and a simple way to define some functions on the object. This process called *recursion* can be used to define sequences, functions, sets and algorithms.

6.4.1 Recurrence Relations

A *recurrence relation* defines a sequence based on a rule that provides the next term of the sequence as a function of the previous terms. For example,

$$a_n = a_{n-1} + 2$$

is a recurrence relation and has the sequence 1, 3, 5, 7,... for $n = 1,...$, which is the set of positive odd integers. Given a recurrence relation, we will attempt to find a closed formula for it. The basic method to apply is to guess the solution and prove it by induction. Consider the following recurrence relation,

$$P_n = 2 \cdot P_{n-1} + 1 \text{ with } P_0 = 0$$

We guess the solution as $2^n - 1$ and prove it by induction.

- *Basis step*: $P_0 = 2^0 - 1 = 0$ is valid.
- *Inductive step*: Assume P_{n-1} is valid, then,

$$P_n = 2 \cdot P_{n-1} + 1 = 2(2^{n-1} - 1) + 1 = 2^n - 1$$

$\square$

Example 6.4.1 Let the recurrence relation be defined as below.

$$P(n) = \begin{cases} 0 & if \ n = 0 \\ 5 \cdot P(n-1) + 1 & if \ n > 0 \end{cases}$$

Prove that $P(n) = \frac{5^n - 1}{4}$ for all $n \geq 0$.
Solution: We will use induction for proof. Let us check the base case; $P(0) = (5^0 - 1)/4 = 0$ which shows that the base case holds. Assuming the recurrence relation is true for $(k-1)$ yields,

$$P(k-1) = (5^{k-1} - 1)/4$$

Substitution for $P(k)$ is as follows.

$$P(k) = 5 \cdot P(k-1) + 1 \qquad \text{by the definition of the function}$$

$$= 5 \cdot \frac{5^{k-1} - 1}{4} + 1 \qquad \text{by the induction hypothesis}$$

$$= \frac{5^k - 5}{4} + 1$$

$$= \frac{5^k - 5 + 4}{4}$$

$$= \frac{5^k - 1}{4}$$

which is $P(k)$ in the recurrence formula. Note that instead of $P(k) \rightarrow P(k+1)$, we used the equivalent conditional statement $P(k-1) \rightarrow P(k)$. $\square$

6.4.2 Recursively Defined Functions

A *recursive function* is defined in terms of itself. In order to specify a recursive function on the set of nonnegative integers, we need to specify the value of the function $f(0)$ for $n = 0$, and provide a rule that states the value of the function $f(n + 1)$ in terms of the value of the function for $i \leq n$. For example $f(n) = a^n$ can be defined recursively as $a^n = a \cdot a^{n-1}$.

An arithmetic sequence can be recursively defined as $a_n = a_{n-1} + d$ with the formula $a_n = a + (n - 1)d$ as noted before. A geometric sequence is defined as $a_n = ra_{n-1}$ with the formula $a_n = ar^n$.

Consider finding the sum of the first n numbers as a function $P : \mathbb{N} \rightarrow \mathbb{Z}$ which inputs a natural number and returns an integer. This function can be defined recursively as follows:

$$P(n) = \begin{cases} 1 & \text{if } n = 1 \\ n + P(n - 1) & \text{if } n > 1 \end{cases}$$

Let us find $P(4)$ using this formula by replacing $P(n)$ with $P(n - 1) + n$ at each step starting from the top yielding the following:

$$\begin{aligned} P(4) &= 4 + P(3) \\ &= 4 + 3 + P(2) \\ &= 4 + 3 + 2 + P(1) \\ &= 4 + 3 + 2 + 1 \\ &= 10 \end{aligned}$$

Note that we need a *base case* $(n = 1)$ to stop replacing the recursive function with a lower value of n for this top-down approach in solving the recursive function.

Example 6.4.2 Define $P(n) = 2^n$ recursively and work out $P(5)$ using top-down approach with this definition.
Solution: $P(n)$ can be defined recursively as follows with the base case $2^1 = 2$

$$P(n) = \begin{cases} 2 & \text{if } n = 1 \\ 2 \cdot P(n - 1) & \text{if } n > 1 \end{cases}$$

$P(5)$ can now be calculated as follows.

$$\begin{aligned} P(5) &= 2 \cdot P(4) \\ &= 2 \cdot 2 \cdot + P(3) \\ &= 2 \cdot 2 \cdot 2 \cdot P(2) \\ &= 2 \cdot 2 \cdot 2 \cdot 2 \cdot P(1) \\ &= 2 \cdot 2 \cdot 2 \cdot 2 \cdot 2 \\ &= 32 \end{aligned}$$

We may be given a recursive function and aim at finding the closed form of this function. In this case, we can list the first values of the function and guess a solution by looking at the values. We can then apply the principle of mathematical induction to prove that the guessed function is correct. For example, let the recursive function $P(n)$ be as follows:

$$P(n) = \begin{cases} 0 & \text{if } n = 1 \\ P(n-1) + 2n - 1 & \text{if } n > 1 \end{cases}$$

Listing of the values of $P(n)$ for $n = 1, 2, 3$ and 4 yields the following:

$$P(1) = 0$$
$$P(2) = 3$$
$$P(3) = 8$$
$$P(4) = 15$$
$$P(5) = 24$$

Based on these values, $f(n) = n^2 - 1$ is our guess. Let us implement induction on $f(n)$; as the basis step, $f(1) = 0$ and recursion gives,

$$
\begin{aligned}
P(k) &= P(k-1) + 2k - 1 && \text{by the definition of the function} \\
&= (k-1)^2 - 1 + 2k - 1 && \text{by the inductive hypothesis} \\
&= k^2 - 2k + 1 - 1 + 2k - 1 \\
&= k^2 - 1
\end{aligned}
$$

which is the expression for $P(k)$ in the recursive definition of the function. □

6.4.3 Recursive Algorithms

A recursive algorithm is a procedure that calls itself with different parameters at each run. It has a base case where returning from the called procedures begin. Let us consider finding the factorial of a positive integer using a recursive procedure. The product of positive integers from 1 to n is called n factorial denoted by $n!$ as noted. This function may be expressed as,

$$n! = n(n-1)(n-2) \cdots 1$$

Factorial of 1 is defined as 1 and hence, this function is defined for the nonnegative integers, also providing a base case. We can write this recursive function as follows,

$$P(n) = \begin{cases} 1 & \text{if } n = 1 \\ n \cdot (n-1)! & \text{if } n > 1 \end{cases}$$

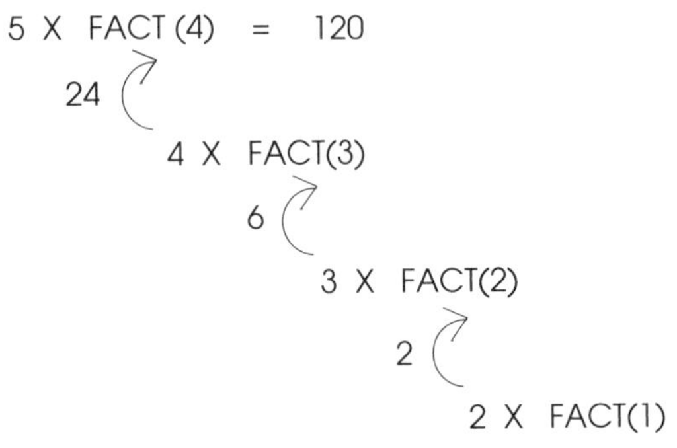

Fig. 6.1 Execution steps of the recursive *Factorial* procedure

$$5 \times FACT(4) = 120$$

$$24$$

$$4 \times FACT(3)$$

$$6$$

$$3 \times FACT(2)$$

$$2$$

$$2 \times FACT(1)$$

We can form a recursive procedure as shown in Algorithm 6.1 which inputs an integer n and calls itself by decreasing the value of n at each call. The returning point from the nested calls to procedures is when $n = 1$ and then the returned value are multiplied to provide $n!$ in the end when the procedure is completed.

Algorithm 6.1 *Recursive Factorial*

1: **procedure** FACT(n:integer)
2:    **if** $n == 1$ **then**
3:        **return**(1)
4:    **else**
5:        **return**($n \cdot$ FACT($n - 1$))
6:    **end if**
7: **end procedure**

The calling sequence of the $FACT$ procedure for $n = 5$ is depicted in Fig. 6.1. The arrows display the returned values from the procedure and the final return provides the caller of this procedure with the value of 120.

Analysis of Recurrence Algorithms

Analysis of a recursive algorithm is needed to determine its time complexity. Our main concern is the number of recursive calls made in the analysis of the algorithm. Let us consider the recursive factorial algorithm with the following number of calls in relation to the input number n.

$$f(1) = 1$$
$$f(2) = 1 + f(1)$$
$$f(3) = 1 + f(2)$$
$$\cdots$$
$$f(n - 1) = 1 + f(n - 2)$$
$$f(n) = 1 + f(n - 1)$$

We can see that $f(n) = 1 + f(n-1)$ for all $n > 1$. By substitution, $f(n) = 1 + f(n-1) = 1 + 1 + f(n-2)$, thus, we can observe the pattern as,

$$f(n) = k + f(n-k)$$

The recursion stops when $n = 1$, thus we can set $n - k = 1$ to get $k = n - 1$. Substituting in the above equation for $f(n)$ yields,

$$f(n) = n - 1 + f(1) = n - 1 + 1 = n$$

Therefore, there is a total of n recursive calls made and the running time of the recursive factorial algorithm is $O(n)$.

Example 6.4.3 Provide a recursive procedure that finds the sum of two integers a and b.

Solution: Instead of simply adding these two numbers, we will define the *Sum* procedure recursively as below. Here, we assume that the return from the base case is when the second number equals 0 and the returned value is the first number. Then we add 1 to each returned value as many times as the second number to find the sum.

$$sum(a, b) = \begin{cases} a & \text{if } b = 0 \\ 1 + Sum(a, b-1) & \text{if } b > 0 \end{cases}$$

We can now form the recursive procedure sum based on the above definition as shown in Algorithm 6.2.

Algorithm 6.2 *Recursive Addition*

1: **procedure** SUM(n:integer)
2:     **if** $b == 0$ **then**
3:         **return**(a)
4:     **else**
5:         **return**($1 + $ SUM($a, b-1$))
6:     **end if**
7: **end procedure**

The working of this algorithm is displayed in Fig. 6.2 for input values $a = 3$ and $b = 5$ with returned values from the recursive calls shown next to arrows.

Example 6.4.4 Provide a recursive procedure that finds the bth power of a positive integer a.

Solution: We will define the recursion with the procedure *Power* as below. This procedure will call itself each time with a decreasing exponent until the exponent reaches 0 and returnings start from that point to the callers.

$$Power(a, b) = \begin{cases} 1 & \text{if } b = 0 \\ a \cdot Power(a, b-1) & \text{if } b > 0 \end{cases}$$

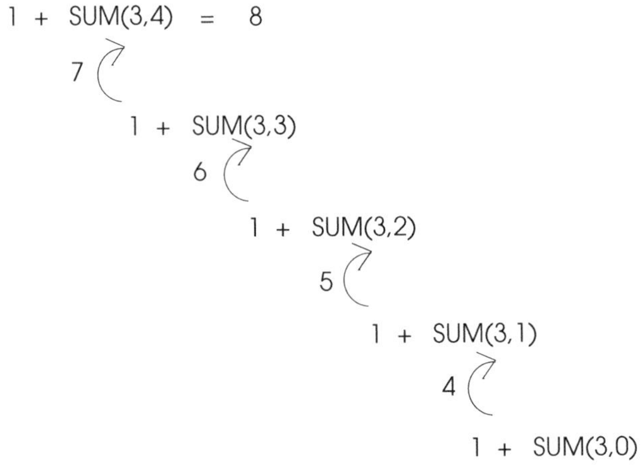

$$1 + \text{SUM}(3,4) \quad = \quad 8$$

$$7$$

$$1 + \text{SUM}(3,3)$$

$$6$$

$$1 + \text{SUM}(3,2)$$

$$5$$

$$1 + \text{SUM}(3,1)$$

$$4$$

$$1 + \text{SUM}(3,0)$$

Fig. 6.2 Execution steps of the recursive *Sum* procedure

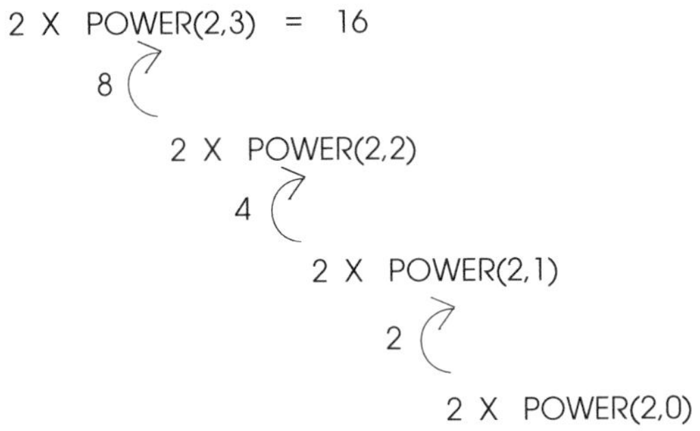

Fig. 6.3 Execution steps of the recursive *Power* procedure

$$2 \times \text{POWER}(2,3) \quad = \quad 16$$

$$8$$

$$2 \times \text{POWER}(2,2)$$

$$4$$

$$2 \times \text{POWER}(2,1)$$

$$2$$

$$2 \times \text{POWER}(2,0)$$

We can now form the recursive procedure *Power* based on the above definition as shown in Algorithm 6.3.

Algorithm 6.3 *Recursive Power*

1: **procedure** POWER(n:integer)
2:    **if** $b == 0$ **then**
3:        **return**(1)
4:    **else**
5:        **return**($a \cdot$ POWER($a, b - 1$))
6:    **end if**
7: **end procedure**

The working of this algorithm is displayed in Fig. 6.3 for input values $a = 2$ and $b = 4$ to yield $2^4 = 16$.

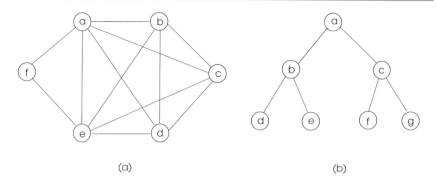

(a)   (b)

Fig. 6.4 **a** A graph. **b** A binary tree

6.4.4 Recursively Defined Sets

Sets can be defined recursively, similar to the definition of recursive functions. Thus, a recursively defined set similarly has a basis step and a recursive step.

Example 6.4.5 Let the set $\Sigma = \{a, b, c, d\}$ and Σ^* set of all strings containing symbols in Σ. The recursive definition of Σ^* can be made as below.

$$\Sigma^* = \begin{cases} \text{empty string } \lambda \in \Sigma^* & \text{if } \Sigma^* \text{ has no elements} \\ wx \in \Sigma^* & \text{if } w \in \Sigma^* \text{ and } x \in \Sigma \end{cases}$$

The basis step states that the empty string is an element of Σ^* and the recursive step states that new strings can produced by adding a symbol from Σ to the end of strings in Σ^*. For example, let the set Σ be $\{p, q, r\}$ and $ppqqr \in \Sigma^*$. Then, $ppqqrp \in \Sigma^*$ with p added.

A *graph* consists of vertices and edges between its vertices. A graph with vertices a, b, c, d, e, f is shown in Fig. 6.4a. A *tree* is a graph with no cycles, in other words, starting from a vertex and traversing through the edges we can not arrive at the starting vertex. A *rooted tree* is an acyclic graph with a distinct vertex called the *root*. Any vertex in a *binary tree* has at most 2 children. A binary tree with a root vertex a is depicted in Fig. 6.4b. We will have a more detailed analysis of graphs in Part II.

Example 6.4.6 A rooted tree can be defined recursively as follows.

- *Basis Step*: A single vertex r is a rooted tree.
- *Recursive Step*: Let $T_1, T_2, ..., T_n$ be disjoint rooted trees with roots $r_1, r_2, ..., r_n$. Then, the rooted tree T with root r can be formed by adding an edge from r to each of the vertices $r_1, r_2, ..., r_n$ as shown in Fig. 6.5.

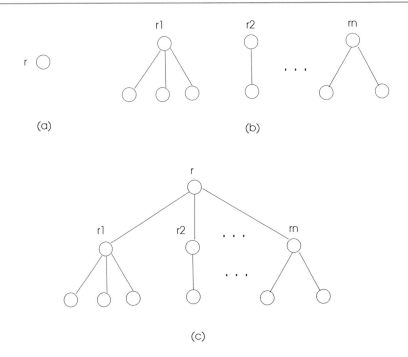

Fig. 6.5 A recursively defined tree, **a** basis step, **b** disjoint trees, **c** is the recursively defined tree

6.5 Structural Induction

Structural induction is a variation of induction over recursively defined sets. This form of induction is commonly used for non-numerical objects where an object can be expressed recursively in terms of smaller objects. It involves two basic steps of induction modified for sets as follows.

- *Basis step*: We need to show that the result holds for all elements stated in the basis step of the recursive definition of the set.
- *Inductive step*: New elements of the set under consideration are formed using the recursive step of the definition and we need to show that if the elements used to construct new elements obeys the proposition, then the new elements also obey the stated recursive step of the definition.

Example 6.5.1 A set S is defined as follows.

- Base elements: 6, 15 $\in S$
- Recursively formed elements: if $a, b \in S$, then $a + b \in S$

Prove that every $n \in S$ generated using the recursive rule is divisible by 3.

Proof We do a simple check, 6 and 15 are both divisible by 3 meaning 3 divides both without a remainder. Numbers 21, 27, 42 etc. can be generated from the base elements and 3 divides all. The structural induction proof follows.

- *Basis step*: 3 divides 6 and 3 divides 15.
- *Inductive step*: Assume $P(a)$ and $P(b)$ are true for some integers a and b in S, we need to show $P(a+b)$ is true. Based on inductive hypothesis, $a = 3x$ and $b = 3y$ for some integers x and y. Thus, $a + b = 3x + 3y = 3(x + y)$ and therefore 3 divides $(a + b)$. □

Example 6.5.2 Any rooted tree with n nodes has $n - 1$ edges.

Proof We need to apply the basis and inductive steps as follows.

- *Basis step*: Let $n = 1$ when the tree has one node. In this case, the tree has no edges.
- *Inductive step*: Let T be a tree with a root r that has k children $c_1, ..., c_k$ each being the root of subtrees $T_1, ..., T_k$ as shown in Fig. 6.6.
  Let the number of vertices in T_i be n_i and edges in T_i be m_i. We can now assume $m_i = n_i - 1$ by the induction hypothesis. There exists k edges from the root r to all of its children and the following can be stated,

$$m = k + \sum_{i=1}^{k} m_i$$

$$= \sum_{i=1}^{k} (1 + m_i)$$

$$= \sum_{i=1}^{k} (1 + (n_i - 1)) \qquad \text{by the inductive hypothesis}$$

$$= \sum_{i=1}^{k} n_i$$

$$= n - 1 \qquad \text{number of nodes except the root}$$

□

Example 6.5.3 Prove that every complete binary tree with n leaves has $n - 1$ internal nodes.

Proof We need to apply the basis and inductive steps as follows.

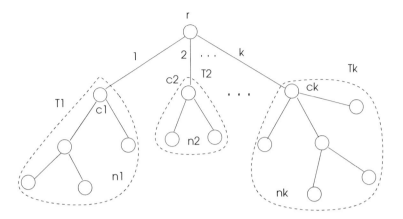

Fig. 6.6 A tree structure

- *Basis step*: Consider the case when $n = 1$ where the tree consists of one node. The tree has one leaf node and $n - 1 = 0$ internal nodes.
- *Inductive step*: Assume a tree that has a root and two subtrees. Let k and m be the number of leaves in the two subtrees, then $k + m = n$. The number of internal nodes using the inductive hypothesis is $(k - 1) + (m - 1) = k + m - 2$. Adding one to this result to count the root, we have $k + m - 1 = n - 1$ as the total.   □

6.6 Review Questions

1. Define an arithmetic sequence.
2. What is the difference between an arithmetic sequence and an arithmetic series?
3. Define a geometric sequence.
4. What is the summation formula for an arithmetic sequence with the first term a_1 and the last term a_n?
5. What is the summation formula for a geometric sequence with a common ratio r?
6. What are the main steps of weak induction?
7. What is the difference between the weak induction and strong induction?
8. What is a recurrence relation?
9. What is a recursively defined function?
10. What are the properties of a recursively defined set?
11. Give an example of a recursive function.
12. What is the base case of a recursive function and how is it defined?
13. What is structural induction and how does it differ from the first and second principles of induction?

6.7 Chapter Notes

We reviewed sequences and summation in the first part of this chapter. A sequence is an ordered list of real numbers and summation is the process of summing the first specified number of elements of a sequence. A term of an arithmetic sequence is formed by adding a constant difference value to its preceding term and a term of a geometric sequence is formed by multiplying its preceding term with a common ratio.

In the second part, we described a powerful method called induction. This method consists of the basis step and the inductive step. Three forms of induction are the weak induction (First Principle of Induction), strong induction (Second Principle of Induction) and the structural induction. The basis step is proven for the base case and the inductive step is to show that $P(k+1)$ holds when $P(k)$ is true in the first method. Strong induction based proofs assume all of the statements $(P(1) \wedge P(2)...P(k))$ are true to show $P(k + 1)$ is true. Structural induction is mainly used for proofs involving recursively defined sets.

Lastly we reviewed recursion which is the process of defining an object in terms of itself. We can have recurrence relations, recursively defined functions, recursively defined sets and recursive algorithms. Recursive algorithms provide powerful and simple ways of performing some difficult computational tasks.

Exercises

1. Find a recurrence relation for the sum $S(n) = 1^2 + 2^2 + 3^2 + ... + n^2$.
2. Give a recursive definition of the following:

   a. The set of positive integers divisible by 7.
   b. The set of positive integers that are multiples of 3.
   c. The set of positive integers congruent to 5 mod 2
   d. The set of integers that are powers of 5.

3. Given the following recurrence relation, compute $P(1)$, $P(2)$, $P(3)$ and $P(4)$.

$$P(n) = \begin{cases} 0 & \text{if } n = 0 \\ (P(n-1))^2 + n & \text{if } n > 0 \end{cases}$$

4. Let A be a finite set with n elements. Find a recurrence relation for the number of elements in the power set (A).
5. Show that the sum of first n even numbers starting from 2 is $n(n+1)$ by induction.
6. Show by induction that $2n + 8$ is even for $n \in \mathbb{Z}$.
7. Show that $\sum_{i=1}^{n} 4i - 2 = 2n^2$ using induction.
8. Show by induction that $n^3 + 2n$ is divisible by 3.
9. Show that $n^3 \equiv n \pmod{3}$ by induction.

10. Prove by induction that

$$1^3 + 2^3 + \ldots + n^3 = \frac{n^2(n+1)^2}{4}$$

11. Compute $\displaystyle\sum_{i=0}^{n-1} 4 \cdot 2^i$

12. Show that the following equation showing the geometric series holds by induction,

$$\sum_{k=0}^{n-1} r^k = \frac{r^n - 1}{r - 1}$$

13. Show that for all integers $n \geq 3$, $2n + 1 < 2^n$ by induction.
14. Write a recursive procedure that finds the product of two positive integers a and b. Use the fact that $a \cdot b$ is a added b times to itself.
15. Show that the cardinality of a set A is 2^n using the induction method.
16. Use strong induction principle to show that the nth Fibonacci number is,

$$F(n) = \frac{\alpha^n - \beta^n}{\alpha - \beta}$$

where

$$\alpha = \frac{1 + \sqrt{5}}{2} \quad \text{and} \quad \beta = \frac{1 - \sqrt{5}}{2}.$$

Introduction to Number Theory

7

Number theory is the study of mostly integers and their relations. This branch of mathematics is old and was often thought to have trivial application areas. However, recent research in cryptography started a renewed interest in number theory which resulted in the design of encryption algorithms that form the basis of secure transactions over the Internet. In this chapter, we will first review basics of number theory, namely; divisibility, modularity, prime numbers and conclude with introduction to cryptography.

7.1 Basics

The set of numbers were classified as follows.

- $\mathbb{R}$: Real numbers.
- $\mathbb{Z}$: Integers.
- $\mathbb{Q}$: Rational numbers.
- $\mathbb{N}$: Positive integers (natural numbers)

$\mathbb{N}^+$ denotes positive integers excluding 0 and $\mathbb{R}$-$\mathbb{Q}$ is the set of irrational numbers. An integer $n \in \mathbb{Z}$ that can be written as $n = 2m + 1$ for some integer m is called an *odd integer*. Similarly, $n \in \mathbb{Z}$ that can be specified as $n = 2m$ for some integer m is an even integer. We have used properties of odd and even numbers for proof examples in Chap. 2. For example, we can state that if both of the integers a and b are odd, their product ab is also odd. In order to prove such propositions, we can substitute $a = 2n + 1$ and $b = 2m + 1$ for some integers m and n. Then $ab = (2n+1)(2m+1) = 2nm + 2n + 2m + 1$ which is $2(2m+n) + 1$. Substituting $k = 2m + 1$, we have $ab = 2k + 1$ which is an odd integer.

© Springer Nature Switzerland AG 2021
K. Erciyes, *Discrete Mathematics and Graph Theory*, Undergraduate Topics in Computer Science, https://doi.org/10.1007/978-3-030-61115-6_7

7.2 Division

The integers form the set $\mathbb{Z} = \{\ldots, -2, -1, 0, 1, 2, 3, \ldots\}$ and the natural number set $N = \{0, 1, 2, 3, \ldots\}$ as noted. One of the main operations done with integers is division defined below.

Definition 7.1 (*division*) Let a and b be two integers in $\mathbb{Z}$. The integer b *divides* a if there exists an integer c such that $c = \frac{a}{b}$. The integers c is called the *quotient*, b the *divisor* or *factor* of a and a the *dividend*. The division is denoted by $b \mid c$ if b divides a, otherwise $b \nmid c$, meaning b does not divide a.

The following properties of division are observed for integers $a, b, c \in \mathbb{Z}$.

1. If $a \neq 0$, then $a \mid 0$.
2. $1 \mid a$.
3. If $a \mid b$ and $b \mid c$, then $a \mid c$. For example, $2 \mid 8$ and $8 \mid 64$, then $2 \mid 64$.
4. If $a \mid b$ and $a \mid c$, then $a \mid (b + c)$. For example, $3 \mid 9$ and $3 \mid 12$, then $3 \mid 21$.
5. If $a \mid b$, then $a \mid (bc)$. For example, $5 \mid 15$, then $5 \mid 45$ for $c = 3$.
6. If $ab \mid b$, then $a \mid c$ and $b \mid c$. For example, $6 \mid 42$, then $2 \mid 42$ and $3 \mid 42$.
7. If $a \mid b$ and $b \mid a$, then $a = b$.
8. If $c \mid a$ and $c \mid b$, then $c \mid (ma + nb)$ $\forall m, n \in \mathbb{Z}^{+}$. For example, $3 \mid 12$ and $3 \mid 18$, then, $3 \mid ((4 \cdot 12) + (5 \cdot 18)) = 3 \mid (48 + 90) = 3 \mid 138$.

These properties may be proven simply by using the definition of divisibility. For example, let us prove the fourth property. If $a \mid b$ then $b = ak$ for some integer k and similarly $a \mid c$ means $c = am$ for some integer m. Then, adding both sides of these two equations yields $b + c = a(k + m)$, thus, $a \mid (b + c)$. $\square$

Proving the last property can be done as follows. If $c \mid a$, then $a = kc$ for some integer k and similarly, $b = pc$ for some integer p since $c \mid a$. Then, substitution yields $ma + nb = mkc + npc = (mk + np)c$, thus, $c \mid (ma + nb)$. $\square$

Example 7.2.1 Show that for all $n \geq 0$, $4 \mid (5^n - 1)$.
Solution: We use induction, $4 \mid (5^0 - 1)$ is true in the basis step. The inductive step is as follows.

$$5^k - 1 = 4m \ \text{ for some integer } m \text{ since } 4 \mid (5^n - 1)$$
$$5(5^k - 1) = 5 \cdot 4m$$
$$5^{k+1} - 5 = 20m$$
$$5^{k+1} - 1 = 20m + 4$$
$$= 4(5m + 1) = 4q \text{ where } q = 5m + 1$$

Therefore, $4 \mid 5^{k+1} - 1$. $\square$

Theorem 7 (division algorithm) *For every two integers a and b, there exists unique integers q and r such that,*

$$a = bq + r, 0 \leq r < b$$

Integer q is the quotient as noted and the integer r is called the *remainder* of a divided by b. For example, dividing 23 by 5 results in quotient $q = 4$ and remainder $r = 3$. We can then write $23 = 4 \cdot 5 + 3$.

Using the properties of division, we can have a procedure that finds the quotient and remainder when integer a is divided by integer b as shown in Algorithm 7.1. We need to subtract b from as many times as q until $r < b$. For example, to find 17 divided by 3, 3 is subtracted 5 times which is q and we stop when the remainder 2 is less than 3. The iterated values of a will be 17, 14, 11, 8, 5 and 2 and r will be similar. The q values will be $1, \ldots, 5$. The running time of this algorithm is simply $O(q)$ and it finishes when $r < b$.

Algorithm 7.1 *Division Algorithm*

1: **procedure** DIVISION(a, b: integers)
2:    $r \leftarrow b$
3:    $q \leftarrow 0$
4:    **while** $r \geq b$ **do**
5:       $q \leftarrow q + 1$
6:       $r \leftarrow a - b$
7:       $a \leftarrow r$
8:    **end while**
9:    **return** q, r
10: **end procedure**

7.3 Greatest Common Divisor

Given two integers a and b, an integer d is a common divisor of a and b if $d \mid a$ and $d \mid b$. We can now define the greatest common divisor of two integers as follows.

Definition 7.2 (*greatest common divisor*) Let a and b be two integers where a and b are not both 0. The *greatest common divisor* of a and b, denoted by $\gcd(a, b)$, is the greatest integer d such that $d \mid a$ and $d \mid b$.

The *gcd* c of two integers a and b is always positive. By definition, c is a common divisor of a and b. Consider the case $c < 0$, then $-c$ is also a divisor of a and b and $-c$ is positive, thus, $-c > c$. The $\gcd(12,36)$ is 12 because 12 is the largest integer that divides both. Similarly, gcd(18,27) is 9. Some important properties of greatest common divisor of two numbers $a, b \in \mathbb{Z}$ may be stated as follows.

1. $\gcd(a, b) = \gcd(b, a)$.
2. $\gcd(a, b) = \gcd(a, -b) = \gcd(-a, b) = \gcd(-a, -b)$.
3. If $a \mid b$ then $\gcd(a, b) = |a|$. For example, $gcd(12, 48) = 12$.
4. If a and b are not both 0, then every common divisor of a and b divides $\gcd(a, b)$.
5. $\gcd(a, b) = \gcd(b, a - mb)$ for all $m \in \mathbb{Z}$. For example, $gcd(42, 12) = \gcd(12, (42 - 3 \cdot 12)) = gcd(12, 6) = 6$.

7.3.1 Euclid's Algorithm

Euclid provided a simple algorithm to find *gcd* of two integers. It is based on the following theorem.

Theorem 8 *Let a and b be two integers such that a > b > 0 and r be the remainder of the division of a by b. Then gcd(a, b) = gcd(b, r).*

Proof Let $a = bq + r$, thus $r = a - bq$. Using the 5th property above,

$$\gcd(a, b) = \gcd(b, a - bq) = \gcd(b, r)$$

□

Based on Theorem 8, Euclid provided a method to find the *gcd* of two integers consisting of the following steps. When the remainder is equal to 0, the parameter a is the *gcd* of the two input numbers a and b.

1. Let $x \leftarrow \max(a, b)$, $y \leftarrow \min(a, b)$
2. **Repeat**
3. $x = qy + r$ (find largest q that satisfies this equation)
4. $x \leftarrow y$
5. $y \leftarrow r$
6. **Until** $r = 0$

Example 7.3.1 Use Euclid's algorithm to find *gcd* of 91 and 26.
Solution: The iterations of the algorithm yields the following:

$$91 = 3 \times 26 + 13$$
$$26 = 2 \times 13 + 0$$

Thus, the returned *gcd* value is 13.   □

Example 7.3.2 Use Euclidean algorithm to find gcd(336, 140).
Solution: The iterations of the algorithm yields the following.

$$336 = 2 \times 140 + 56$$
$$140 = 2 \times 56 + 28$$
$$56 = 2 \times 28 + 0$$

The returned *gcd* value is 28.   □

Recursive Formulation

Let us attempt to have a recursive version of Euler's algorithm. The recursion based on Theorem 8 can be written as below by letting r to be the remainder of the division of a by b all being integers.

$$\text{gcd(a,b)} = \begin{cases} a & \text{if } b = 0 \\ \text{gcd}(b, a \bmod b) & \text{if } (b > 0) \end{cases}$$

The recursive algorithm is depicted in Algorithm 7.2.

Algorithm 7.2 *Euclid's Recursive Algorithm*

1: **procedure** EUCLID_REC(a, b: positive integers)
2:     ($r \leftarrow a \bmod b$)
3:     **if** $r = 0$ **then**
4:         **return** (b)
5:     **end if**
6:     **return** EUCLID_REC(b, r)
7: **end procedure**

7.3.2   Least Common Multiple

Definition 7.3 (*least common multiple*) Let a and b be two integers in $\mathbb{N}^+$. The *least common multiple* (LCM) of a and b, denoted by lcm(a, b), is the smallest integer $d \in \mathbb{N}^+$ such that $a \mid d$ and $b \mid d$.

Example 7.3.3 The LCM of 27 and 18 is 54 since is the smallest integer that both divides. Similarly, lcm(12, 15) is 60.

Some important properties of LCM may be stated for integers $a, b \in \mathbb{Z}$ as follows.

- lcm(a, b) = lcm(b, a)
- lcm(a, b) = lcm($a, -b$) = lcm($-a, b$) = lcm($-a, -b$)
- lcm(ka, kb)=$|k|$ lcm(a, b). For example, lcm(24,36)=4 · lcm(6,9)= 4 · 3 = 12

- If $a \mid b$ then $\mathrm{lcm}(a, b) = |b|$. For example; $2 \mid 8$, then $\mathrm{lcm}(2, 8) = 2$.

Given two integers a and b, we can set the greater of a and b to a variable g and increment g until it can be divided by both. For example, let $a = 9$, $b = 12$, thus, $g = 12$. It can not be divided by both, therefore we continue increasing g up to 36 when it can be divided by both. Thus $\mathrm{lcm}(9, 12) = 36$. An algorithm formed based on this procedure is depicted in Algorithm. 7.3.

Algorithm 7.3 *LCM*

1: **procedure** LCM(a, b: positive integers)
2:     **if** $g \geq a$ **then**
3:         $g \leftarrow a$
4:     **else**
5:         $g \leftarrow b$
6:     **end if**
7:     **while** *true* **do**
8:         **if** $g \bmod a = 0$ and $g \bmod a = 0$ **then**
9:             **return** g
10:         **end if**
11:         $g \leftarrow g + 1$
12:     **end while**
13: **end procedure**

Instead of incrementing the larger of the two integers, we can multiply the greater one (g) iteratively with $1, 2, \ldots$ up to smaller one (s) until it can be divided by both as shown in Algorithm 7.4. The upper limit of incrementing the index of the *for* loop is the smaller integer since we will have found a multiple of both integers when greater is equal to greater times the smaller.

Algorithm 7.4 *LCM2*

1: **procedure** LCM(a, b: positive integers)
2:     **if** $g \geq a$ **then**
3:         $g \leftarrow a, s \leftarrow b$
4:     **else**
5:         $g \leftarrow b, s \leftarrow a$
6:     **end if**
7:     **for** $i = 1$ to s **do**
8:         $lcm \leftarrow g \cdot i$
9:         **if** $lcm \bmod s = 0$ **then**
10:             **return** lcm
11:         **end if**
12:     **end for**
13: **end procedure**

Another method to find the LCM of two integers is to find their *gcd* and divide the product of two numbers by their *gcd* since,

$$lcm(a, b) \cdot gcd(a, b) = a \cdot b$$

For example, gcd(18, 24) = 6 and lcm(18, 24) = (18·24)/6 = 72. Thus, finding *gcd* of two integers using Euclid's algorithm also provides LCM of them.

7.4 Prime Numbers

Prime numbers are an important class of integers and they have important applications in cryptography as we will see.

Definition 7.4 (*prime*) A *prime number*, or simply a *prime*, is a positive integer $p \geq 2$ divisible by only 1 and itself. A positive integer $p \geq 2$ that is not prime is called a *composite*.

Definition 7.5 (*coprime*) Two integers with a *gcd* of 1 are called *relatively prime numbers* or *coprimes*.

Example 7.4.1 Let us consider 26 and 33. Factors of 26 are 1, 2, 13 and 33 has 1, 3 and 11 as factors. Their only common factor is 1, therefore these numbers are coprimes. Note that two relatively prime numbers need not be primes as in this example.

Example 7.4.2 Show that any two consecutive integers n and $n + 1$ are coprimes. *Solution*: Let $d = \gcd(n, n + 1)$, that is, there exists and integer $d \neq 1$ that divides both of thes two consecutive numbers. Then $d \mid n$ and $d \mid (n + 1)$ and we can state the following.

$$n = dk \text{ for some integer } k$$
$$n + 1 = dm \text{ for some integer } m$$
$$1 = d(m - k)$$

which means $d \mid 1$, thus, $d \leq 1$. Since d must be greater than or equal to 1, we conclude $d = 1$.   □

Some properties of coprimes for integers $a, b \in \mathbb{Z}$ can be listed as follows [1].

- If $c \mid a$ and a and b are coprimes, then c and b are coprimes.
- If $c \mid ab$ and a and c are coprimes, then $c \mid b$.
- If a and b are coprimes, then $\gcd(a, bc) = \gcd(a, c)$.

We proved the following theorem in Sect. 6.3, called the *fundamental theorem of arithmetic* which shows a factorization of an integer using primes, we will just state it here.

Theorem 9 (prime factor decomposition) *Every positive integer can be expressed as a product of prime numbers.*

Example 7.4.3 Let us consider 38, 56, 77 and 102. These numbers can be factorized into primes as follows: $38 = 2 \cdot 19$; $56 = 2^3 \cdot 7$, $77 = 7 \cdot 11$ and $102 = 2 \cdot 3 \cdot 17$. Note that a prime number cannot be further factorized by the definition of a prime number.

A simple way to find prime factors of a number is to first factorize it to any of its factors and then finding prime factors of these factors. For example, consider 72 which is $8 \cdot 9$, and $8 = 2^3$, $9 = 3^2$. Thus, $72 = 2^3 \cdot 3^2$. Moreover, finding the *gcd* of two integers can be done by finding prime factorizations of these integers and then finding the product of common prime factors with the least power, since these factors should be common to both.

Example 7.4.4 Find gcd(112, 70) using prime factorization.
Solution: Prime factorization of these two numbers are as follows.

$$112 = 2 \times 2 \times 2 \times 2 \times 7 = 2^4 \times 7$$
$$70 = 2 \times 5 \times 7$$

Common prime factors are 2 and 7 and taking the smallest common powers of these factors yield $2 \times 7 = 14$ as the *gcd*. If we had applied Euclid's algorithm to these integers, we would find,

$$112 = 1 \times 70 + 42$$
$$70 = 1 \times 42 + 28$$
$$42 = 1 \times 28 + 14$$
$$28 = 2 \times 14 + 0$$

to yield 14 as the same result.   □

7.4.1 Primality Test

The prime numbers between 1 and 100 are: 2, 3, 5, 7, 11, 13, 17, 19, 23, 29, 31, 37, 41, 43, 47, 53, 59, 61, 67, 71, 73, 79, 83, 89 and 97. As the numbers get bigger, it is more difficult to check whether a given number is prime or not. Finding very large prime numbers is a challenge and there are frequent reports of newly found very large prime numbers [2]. We can have three basic approaches to test whether a given integer n is prime or not.

1. *Brute Force Method*: Test whether any number $2 \leq a < n$ divides n. If no such number is found, n is prime.

Example 7.4.5 We want to test whether 19 is prime. For all numbers $2, \ldots, 18$ none divides 19. Therefore 19 is prime.   □

2. *Use of Prime Factors*: We know by Theorem 9 that every composite number can be written as a product of primes. We can use this fact to establish whether a given number n is prime or not. This way, we need to test only primes less than n.

Example 7.4.6 Let us check whether 23 is prime. For all primes less than 23 which are 2, 3, 5, 7, 11, 13, 17, 19; none divides 23. Therefore 23 is prime. We now have a shorter list to check.   □

3. Our last method is based on the following theorem.

Theorem 10 *Any composite number $n \geq 2$ is divisible by some factor $p \geq \sqrt{n}$ where p is a prime.*

Proof Since n is not a prime, it can be factored as $n = ab$ with $1 < a \leq b < n$. Consider the case where $a > \sqrt{n}$ and $b > \sqrt{n}$. Then, $ab > \sqrt{n}\sqrt{n}$ which is a contradiction. Therefore, n has a divisor d less than $\sqrt{n}$. By Theorem 9, d is either prime or a product of primes. In either case, n has a prime divisor less than $\sqrt{n}$. □

This means primality test of an integer n involves testing whether any prime number less than $\sqrt{n}$ divides it.

Example 7.4.7 Let us consider the primality of 213. We need to test all primes less than $\sqrt{213}$ which is a real number between 14 and 15. Primes less than 14 are 2, 3, 5, 7, 11, 13 and none divides 213. Therefore 213 is prime.   □

As a result, *primality test* of a natural number n involves testing whether a prime divisor less than or equal to $\sqrt{n}$ exists. We can have simple algorithm to test whether a given integer n is prime or not as shown in Algorithm 7.6.

Theorem 11 *There are infinitely many primes.*

Proof The proof is by contradiction. Assume there are a finite number of primes $P = \{p_1, p_2, \ldots, p_k\}$ which are less than or equal to a prime number p_m in this set. Let $n = p_1.p_2....p_k+1$. Hence, n is larger than each p_i and since P covers all primes, n must be composite which means it has a prime divisor. Note that when n is divided by p_i in this set, the remainder is 1, thus, $p_i \nmid n$. Thus, letting q be a prime

---

Algorithm 7.5 *Primality Test*

1: **procedure** PRIMALITY(n: integer)
2:    **for all** primes p up to $\lfloor \sqrt{n} \rfloor$ **do**   ▷ test all primes up to $\sqrt{n}$
3:       **if** $p \mid n$ **then**
4:          **return** *Composite*
5:       **end if**
6:    **end for**
7:    **return** *Prime*
8: **end procedure**

---

factor of n, $q \notin P$. Then, it must be the case that $q > p_m$ which means we have a prime greater than p_m. Therefore, the number of primes is infinite.   □

Example 7.4.8 Show that there is a prime larger than 7 using Theorem 11.
Solution: The primes less than or equal to 7 are 2, 3, 5 and 7. Let $n = 2 \cdot 3 \cdot 5 \cdot 7 + 1 = 211$. Let us check whether 211 is prime or composite; $14 < \sqrt{211} < 15$ and none of the prime numbers less than 14 which are 2, 3, 5, 7, 11, 13 divide 211, thus 211 is prime. If we had found 211 as not prime, we would search a prime factor of 211 that is greater than any of 2, 3, 5 and 7.

Definition 7.6 (*twin primes*) Two primes $a, b \in \mathbb{Z}^+$ with $b = a + 2$ are denoted *twin primes*.

For example, (3, 5), (5, 7), (11, 13), (17, 19) are twin primes. The following conjectures about prime numbers are well-known but are not proven to date.

- *Goldbach's conjecture* states that every even integer greater than 4 can be written as the sum of two primes, these to primes can be the same number. For example, $8 = 3 + 5$, $10 = 3 + 7$ or $10 = 5 + 5$.
- There are infinite twin primes.

7.4.2 The Sieve of Eratosthenes

Eratosthenes of ancient Greece provided a simple method to find all primes up to a given integer n. This method consisted of the following steps.

1. List the integers from 2 to n.
2. Circle the first integer in the list that is not crossed or circled and cross all of its multiples.
3. Repeat until all integers in the list are either circled or crossed.

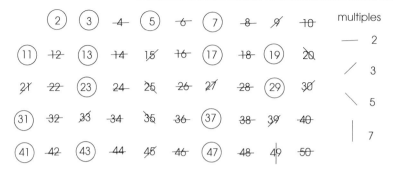

Fig. 7.1 Secure message transfer

Thus, the first integer circled is 2 and all its multiples are crossed. The second integer circled is 3 and all its multiples are also crossed and this process continues when all integers in the range are circles or crossed out. The circled integers are the primes. Operation of this method is depicted in Fig. 7.1 with multiples of 2, 3, 5 and 7 crossed out shown by $-, /,$ and $|$ symbols respectively. The main idea of this method is that if a number is a multiple of some base number, then it is not a prime and can be discarded.

7.5 Congruence

Two integers are congruent modulo n if they have the same remainder when divided by n. An even integer has a remainder of 0 when divided by 2 and an odd integer has 1. The formal definition of congruence is as follows.

Definition 7.7 (*congruence*) Given integers $a, b,$ and $k > 1, a$ is said to be *congruent* to b *modulo* k if $k \mid (a - b)$. This relationship is stated as $a \equiv_k b$, or more commonly, $a \equiv b \pmod{k}$, we will use both notations interchangeably. If these integers are not congruent, this is stated as $a \not\equiv b \pmod{k}$.

Example 7.5.1 $5 \equiv 27 \pmod{2}$ since $5 - 27 = -22$ is divisible by 2. $-7 \equiv -32 \pmod{5}$ since $-7 - (-32) = 25$ is divisible by 5.

Some important arithmetic properties of the congruence relation can be stated for integers as follows.

1. If $a \equiv b \pmod{m}$, then $b \equiv a \pmod{m}$.
2. If $a \equiv b \pmod{m}$ and $b \equiv c \pmod{m}$, then $a \equiv c \pmod{m}$
3. If $a \equiv b \pmod{m}$, then $a + c \equiv b + c \pmod{m}$ and $ac \equiv bc \pmod{m}$.
4. If $a \equiv b \pmod{m}$ and $c \equiv d \pmod{m}$, then $a + c \equiv b + d \pmod{m}$.

5. If $a \equiv b$ (mod m) and $c \equiv d$ (mod m), then $ac \equiv bd$ (mod m).
6. If $a \equiv b$ (mod m) and $m \geq 0$ then $a^k \equiv b^k$ (mod m).

Let us prove the fourth property; $m \mid (a - b)$ and $m \mid (c - d)$ by definition. Then,

$$m \mid ((a - b) + (c - d)) = m \mid (a + c) - (b + d))$$

which means $a + c \equiv b + d$ (mod m) by the definition of congruence.   □

Example 7.5.2 Let us use the 5th property to calculate 351 (mod 6). Since $351 = 13 \cdot 27$,

$$13 \equiv 1 \quad (\text{mod } 6)$$
$$27 \equiv 3 \quad (\text{mod } 6)$$
$$351 \equiv 3 \quad (\text{mod } 6)$$

The set-related properties of the congruence which make it an equivalence relation can be listed as below. An equivalence relation partitions the set under consideration into equivalent classes.

- *Reflexive*: $a \equiv a$ (mod x) for any integer a since $a - a = 0$ is divisible by x.
- *Symmetric*: If $a \equiv b$ (mod x), then $b \equiv a$ (mod x) because the former means $x \nmid (a - b)$. Then, we can say $x \nmid (b - a)$ since $b - a = -(a - b)$.
- *Transitive*: If $a \equiv b$ (mod x) and $b \equiv c$ (mod x), then $a \equiv c$ (mod x). If $(a - b)$ and $(b - c)$ are both divisible by x, then their sum is also divisible by x.

Theorem 12 $a \equiv b$ (mod m) *if and only if* $b = a + mq$ *for some integer* q.

Proof We need to prove both directions of the statement. If $a \equiv b$ (mod m), then $m | (b - a)$ by the definition. Thus, $b - a = mq$ for some integer q. Hence, $b = a + mq$. In the other direction, if $b = a + mq$, then $b - a = mq$ which means $m | (b - a)$ and thus $a \equiv b$ (mod m).   □

We can use Theorem 12 for calculations of congruence values, for example, given $64 \equiv b$ (mod 18), b can be found by subtracting 3×18 from 64 to result in 10. Thus, $64 \equiv 10$ (mod 18). The 6th property stated above provides congruence values for large numbers. For example, $14 \equiv 9$ (mod 5) means we can state $14^2 \equiv 9^2$ (mod 5), that is, $196 \equiv 81$ (mod 5).

Example 7.5.3 Find 5^{17} (mod 7).
Solution:

$$5^2 \equiv_7 4$$
$$(5^2)^2 \equiv_7 16 \equiv_7 2$$
$$(5^4)^2 = 5^8 \equiv_7 4$$
$$(5^8)^2 = 5^{16} \equiv_7 16 \equiv_7 2 \quad \text{now use 4th property,}$$
$$(5^{17}) = 5^{16+1} = 5^{16} \cdot 5 \equiv_7 2 \cdot 5 = 10 \equiv_7 3$$

Example 7.5.4 Find 13^{381} (mod 5).
Solution:

$$13 \equiv_5 3$$
$$13^2 \equiv_5 9 \equiv_5 4$$
$$13^4 \equiv_5 16 \equiv_5 1$$
$$(13^4)^{95} = 1^{95} \equiv_5 1$$
$$(13^{381}) = 13^{380+1} \equiv_5 1 \cdot 13 \equiv_5 3$$

Definition 7.8 (*residue class*) The set $\bar{a} = \{x \in \mathbb{Z} \mid x \equiv a \pmod{m}\}$ of all integers that are congruent modulo m to a is called a *residue class*, or a *congruence class*, modulo m.

For example, $2 \equiv 7 \equiv 12 \equiv \cdots \pmod 5$ is the congruence class 2 (mod 5). For any integer m, the number of residue classes is m. These classes are represented by $0, 1, \ldots, m - 1$.

Definition 7.9 (*linear congruence equation*) Let us consider the following equation,

$$ax \equiv b \pmod{m}$$

with $m \in \mathbb{Z}$ and $0 \le b < m$. This equation is called the *linear congruence equation* and our aim is to determine the set of values for the parameter x.

Let $ax \equiv b \pmod m$ be a linear congruence equation and let $d = \gcd(a, m)$. If $d = 1$ meaning a and m are coprimes, and $d \mid b$, there is exactly one solution to the equation. If $d \nmid b$, there is no solution to the equation and if $d > 1$ and $d \mid b$, there are exactly d solutions; we mean a residue class by a single solution. By definition, $ax \equiv b \pmod m$ means $m \mid (ax - b)$, thus, $ax - b = my$ for some integer y. Therefore,

$$ax - my = b$$

which is a diophantine equation meaning an equation we search integer solutions for x and y. Since d divides both a and m, there is no solution of this equation if $d \nmid b$, therefore linear congruence has no solutions. If $d \mid b$,

$$x = x_0 + \left(\frac{m}{d}\right)t, \quad y = y_0 + \left(\frac{a}{d}\right)t \quad t = 0, 1, .., d - 1 \qquad (7.1)$$

Thus, if we find x_0, we can find the rest of the solutions.

Example 7.5.5 Let $2x \equiv 5 \pmod{6}$. Then, $d = \gcd(2, 6) = 2$ which does not divide 5, thus, there is no solution to x. It can be observed that $2x$ is always an even number and an even number divided by an even number will always give an even remainder so 5 as a remainder will not be possible.

Example 7.5.6 Let $4x \equiv 3 \pmod{7}$. Then, $d = \gcd(4, 7) = 1$ which divides 3, thus, there is only one solution to x. Trial shows $x = 6$, in fact the residue class 6 (mod 7). For example, $x = 13, 20, 27, \ldots$ are all solutions. Testing shows $4 \cdot 13$ (mod 7), $4 \cdot 20$ (mod 7), $4 \cdot 27$ (mod 7) are all congruent to 3 (mod 7).

Example 7.5.7 Let $4x \equiv 2 \pmod{6}$. Then, $d = \gcd(4, 6) = 2$ which divides 2, thus, there are two solutions to x. The first one, x_0, by trial is 2 and the second one x_1 by Eq. ?? is $2 + (6/2) \cdot 1 = 5$. The residue classes are 2 (mod 6) and 5 (mod 6). For example, $x = 2, 8, 14, \ldots$ and $x = 5, 11, 16, \ldots$ are all solutions. Thus, $4 \cdot 2 \equiv 2$ (mod 6), $4 \cdot 8 \equiv 2$ (mod 6), and $4 \cdot 5 \equiv 2$ (mod 6), $4 \cdot 11 \equiv 2$ (mod 6).

Definition 7.10 (*multiplicative inverse*) A multiplicative inverse of an integer a (mod m) exists if and only if $\gcd(a, m) = 1$. This parameter is denoted by a^{-1} where

$$a \cdot a^{-1} = 1 \pmod{m}$$

A simple method to work out the solutions once $\gcd(a, m) = 1$ is determined is to multiply each side of the linear congruence by the multiplicative inverse of a. For example, let the linear congruence be,

$$3x \equiv 5 \pmod{7}$$

The multiplicative index of 3 (mod 7) is 5 since $3 \cdot 5 \equiv 1 \pmod{7}$. Multiplying each side of the above equation by 5 yields,

$$3x \equiv 5 \pmod{7}$$
$$(3 \cdot 5)x \equiv 5 \cdot 5 \pmod{7}$$
$$x \equiv 4 \pmod{7}$$

Thus $x = 4, 11, 18, \ldots$ are all solutions to the given linear congruence.

7.6 Representation of Integers

Decimal system uses integers $0, \ldots, 9$ to represent digits and 10 as the base. Each digit is multiplied by the increasing powers of 10 as we move from right to left to show the expanded representation of a decimal number as shown in below example.

$$5236 = 5 \cdot 10^3 + 2 \cdot 10^2 + 3 \cdot 10^1 + 6 \cdot 10^0$$

Definition 7.11 (*base b representation*) The base b representation of an integer $a \in \mathbb{N}$ is $(a_{n-1}a_{n-2}...a_0)_b$ with integers $0 \le a_i < b$ satisfies the following equation.

$$(a_{n-1}a_{n-2}...a_0) = a_{n-1}b^{n-1} + a_{n-2}b^{n-2} + \cdots + a_1 b + a_0 \qquad (7.2)$$

Binary and hexadecimal representation of integers are commonly used in computer systems as described next.

7.6.1 Binary System

The binary system uses 2 as the base and a binary digit $d \in \{0, 1\}$ is called a *bit*. Data and instructions in a computer are stored as bits in memory. A binary number and its decimal equivalent using Eq. 7.2 is shown below.

$$1101\ 0101 = 1 \cdot 2^7 + 1 \cdot 2^6 + 0 \cdot 2^5 + 1 \cdot 2^4 + 0 \cdot 2^3 + 1 \cdot 2^2 + 0 \cdot 2^1 + 1 \cdot 2^0 = 213.$$

Algorithm 7.6 show, how to convert a binary number to decimal where each digit is multiplied by powers of base.

Algorithm 7.6 *Decimal Conversion*

1: **procedure** DEC_CONVERT(n: number of digits, b: base, a: number to convert)
2:     $d\_val \leftarrow 0$
3:     $b\_ind \leftarrow 1$
4:     **for** $i = 0$ to $n - 1$ **do**
5:         $d\_val \leftarrow d\_val + b\_ind \cdot a(i)$
6:         $b\_ind \leftarrow b\_ind \cdot b$
7:     **end for**
8:     **return** d_val
9: **end procedure**

We can also convert a decimal d number to a given base b by successively dividing d by b and recording the remainders. For example, conversion of 23_{10} to the binary number 10111 is shown in Fig. 7.2. The remainders are shown in circles. The remainder of the first division provides the *least significant bit* of the binary number and the remainders of the divisions at each iteration thereafter provide the digit values for $2^1, 2^2, \ldots$. We stop when the quotient is 0.

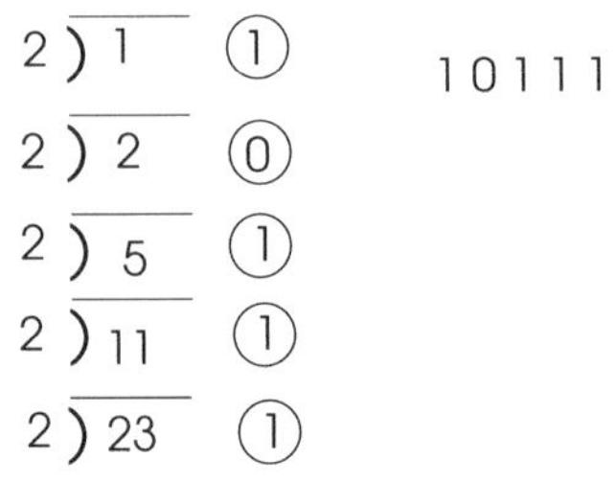

Fig. 7.2 Decimal to binary
conversion

$$2 \,) \,\overline{1} \quad \textcircled{1} \qquad\qquad 1\ 0\ 1\ 1\ 1$$
$$2 \,) \,\overline{2} \quad \textcircled{0}$$
$$2 \,) \,\overline{5} \quad \textcircled{1}$$
$$2 \,) \,\overline{11} \quad \textcircled{1}$$
$$2 \,) \,\overline{23} \quad \textcircled{1}$$

7.6.2 Hexadecimal System

The hexadecimal system uses 1,..,9, A, B, C, D, E, F with symbols A to F representing integers 10–15 with the idea of representing all integers in the range 0, . . . , 15 with a single symbol, thus, the base of this system is 16. An example hexadecimal number with its decimal equivalent is shown below.

$$A2D = 10 \cdot 16^2 + 2 \cdot 16^1 + 13 \cdot 16^0 = 305_{10}$$

Hexadecimal system is commonly used to show the contents of memory locations in a computer. Addition of two hexadecimal numbers can be done in the usual way as with decimal numbers by considering the base as 16 as shown in below example,

$$
\begin{aligned}
5\ A\ B \quad &= 5 \cdot 16^2 + 10 \cdot 16^1 + 11 \cdot 16^0 = 1,451 \\
C\ F\ 4 \quad &= 12 \cdot 16^2 + 15 \cdot 16^1 + 4 \cdot 16^0 = 3,316 \\
+\overline{} \\
1\ 2\ 9\ F \quad &= 3 \cdot 16^3 + 14 \cdot 16^2 + 10 \cdot 16^1 + 15 \cdot 16^0 = 4,767
\end{aligned}
$$

7.7 Introduction to Cryptography

Cryptography is the study of sending and receiving messages in a secure way and *cryptology* is the process of analyzing *cryptosystems* which aim to provide secure communications. Cryptography involves two distinct processes for a message transfer between an entity A and entity B. These entities can be persons, computers but more commonly, a person and a server computer.

- *Encryption*: A message called *plaintext* in original format is converted to a format called *ciphertext* using a *key* by A which is assumed to be not understood by a third party. A key is used to transfer the message to a special format.
- *Decryption*: The encrypted message received by B is decrypted using the key to recover the original message.

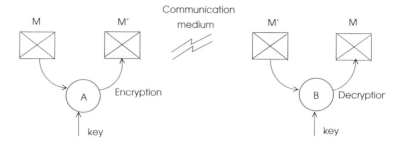

Fig. 7.3 Secure message transfer

Table 7.1 One-time pad method

m	k	$m \oplus k$	$k \oplus (m \oplus k)$
0	0	0	0
0	1	1	0
1	0	1	1
1	1	0	1

This method is depicted in Fig. 7.3. Note that the key function must be decided between A and B prior to message transmission.

Julius Ceasar used a key that consisted of shifting each letter of alphabet 3 positions to the right to communicate with his commanders using the Roman alphabet of 22 letters. The receiving commander would then need to shift each letter received 3 positions back. For example, ATTACK would be transferred DWWDFN using English alphabet. Clearly, the key must be kept secret between the exchangers of messages. Knowing the encryption key provides calculation of the decryption key as in this example, the receiver would need to shift each letter 3 positions back in the alphabetic sequence.

A more reliable private key protocol called *one-time pad* works by the sender of the message *xor*ing each bit of the message with a key known *en priori* by the sender and the receiver. The received message is decrypted by *xor*ing each bit of the message to recover the original message. This method is based on the fact that $k \oplus (m \oplus k) = m$ where k is the key and m is the message, as shown in Table 7.7.

Example 7.7.1 A message and its encrypted version is as follows.

$$\text{message} : 1011\ 0011\ 1101$$
$$\text{key} : 1010\ 0110\ 0111$$
$$\text{encrypted message} : 0001\ 0101\ 1010$$

The receiver performs the following to retrieve the message.

$$\text{encrypted message} : 0001\ 0101\ 1010$$
$$\text{key} : 1010\ 0110\ 0111$$
$$\text{decrypted message} : 1011\ 0011\ 1101$$

The key in this method, however being more difficult to discover than a simple shifting method, can be broken by simply analyzing the frequencies of the codes transmitted by the frequency of letters in the alphabet in use, for example, E is the most common letter in English alphabet followed by A, O and I.

What we have described up to this point is *private key cryptography* characterized by a private secret key known by the sender and receiver of the message. In a rather different approach known as *public key cryptography*, the key of the sender is made public so that it is known by any third party. However, the decryption key is known only by the receiver of the message. Two important public key cryptography protocols in use are described briefly in the next sections.

7.7.1   Diffie-Hellman Protocol

An obvious problem with private key cryptography is keeping the key secret. A and B need to share the secret key prior to communication and if they can do this reliably over the network beforehand, then there is hardly any need for encryption and decryption since they can use whatever was used as protocol for key transfer. Diffie-Hellman (DH) protocol provided calculation of the same key by both parties prior to message transmission without the transfer of the actual key. It consists of the following steps between two entities A and B [3].

1. A selects a secret number a and B selects a secret number b known only to themselves.
2. A performs some computation on her secret number and produces a', B also does some computation on his number b and produces b'.
3. A and B exchange the produced numbers a' and b' which become public.
4. A now performs some computation on a and b' to produce a key K.
5. B performs some computation on b and a' to produce the same key K. The DH algorithm ensures that both parties calculate the same key K.

After this initial process, A and B can use a conventional encryption/decryption protocol. Note that this protocol ensures that any third party observing the message transfers can not produce the key without knowing the value of a or b.

7.7.2   RSA Protocol

This system was proposed by three researches, R. L. Rivest, A. Shamir and L. M. Adleman and hence the name. It is widely used for secure Internet message transfers including secure financial transactions. The RSA cryptology algorithm is basically a public key cryptography method with a public encryption key and a secret decryption key. It consists of three steps:

- *Preparation*: The receiver B prepares public key to be used by all senders to encrypt a message to send to B and it broadcasts this information which can be observed by third parties. This step is performed only once by B.
- *Encryption*: Sender A encrypts her message M using the public key and sends it as ciphertext C.
- *Decryption*: Receiver B converts C back to M using her private key.

Let us review some theory before looking at this protocol in detail.

Definition 7.12 (*Totient number*) The *totient number* of an integer n denoted by $\phi(n)$ is defined as the number of integers less than or equal to n which are coprimes of n. Formally,

$$\phi(n) = |\{a \in \mathbb{Z} : 1 \le a \le n \text{ and } \gcd(a, n) = 1\}|$$

Example 7.7.2 Applying this function to integers 9, 12 and 15 we find, $\phi(9)$=7 since 1, 2, 4, 5, 7, and 8 are coprimes of 9 in this range. Similarly, $\phi(12) = 4$ as 1, 5, 7 and 11 are coprimes of 12 and $\phi(15) = 9$ since 1, 2, 4, 6, 7, 8, 11, 13 and 14 are coprimes of 15 between 1 and 15.

We will now state few results without proving them using this definition.

- If p is prime, then $\phi(p) = p - 1$. This result is simply by considering that all integers up to p are coprimes of p, otherwise p would not be a prime.
- $\phi(p \times q) = \phi(p) \times \phi(q) = (p - 1)(q - 1)$
- *Euler's Theorem*: If a and n are coprimes, then $a^{\phi(n)} \equiv 1 \bmod n$.
- If p is prime, then $a^{p-1} \equiv 1 \bmod p$, $\forall a, 1 \le a \le p - 1$.

Let us first look the steps of the algorithm from the receiver side. Let A be the sender and B the receiver as before.

1. B selects two large primes p and q and computes $n = pq$. Although n will be made public, it is very difficult to factorize p and q from n.
2. She then computes $\phi(n) = (p - 1)(q - 1)$ and chooses an integer e to satisfy $gcd(e, \phi(n)) = 1$. In other words, she selects e to be a coprime of the tuple (e, d) which is made public.

3. She computes a unique number $e, 0 < e < \phi(n)$ such that $e \cdot d = 1 \pmod{\phi(n)}$ which is the key to be used for decryption.

The sender A whenever she wants to send a message $M, 0 \leq M \leq n - 1$, she computes $C = M^n \bmod n$ and transfers C to B. The receiver B then computes $C^d \bmod n$ which is equal to M.

Example 7.7.3 We have the following steps to implement an example scenario:

1. Let us select $= 3$ and $q = 5$. Then $n = 15$ and $\phi(n) = 2 \cdot 4 = 8$.
2. We need to compute e such that $\gcd(e, 8) = 1$, select 11.
3. Now, $11d = 1 \bmod 8$; considering $11d = 8k + 1$, one possible value for d is 3 for $k = 4$.

To send message $M = 7$ for example, sender A encrypts the message using the public key (11,15). The ciphertext $C = 7^{11} \bmod 15 = 13$. The receiver B decrypts the message using the private key (3,15) to recover $M = 13^3 \bmod 15 = 2197 \bmod 15 = 7$.

7.8 Review Questions

1. Define the division of two integers and give an example.
2. What is meant by "a divides b" for two integers a and b.
3. What is the greatest common divisor of two integers a and b?
4. List the main properties of the greatest common divisor of two integers a and b.
5. Describe the steps of Euclid's algorithm to find the greatest common divisor of two integers.
6. What is the least common multiple of two integers a and b?
7. List the main properties of the least common multiple of two integers a and b
8. Define a prime number and give an example.
9. What is the fundamental theorem of arithmetic?
10. Describe the method of Erosthesis to find the primes up to a given number n.
11. Describe the steps of an algorithm to find the primes up to a given number $n4$.
12. Define congruence of two integers and give an example.
13. Describe briefly the Diffie-Hellman protocol of public key cryptography.
14. Describe briefly the RSA protocol of public key cryptography.

7.9 Chapter Notes

We started this chapter by the definition of division of two integers. Greatest common divisor of two integers is the largest integer that divides both of them. The least common multiple of two integers is the smallest integer that is a multiple of both of them. We listed properties of these two functions.

Prime numbers are an important class of integers and they have many applications including cryptography. A prime number can be divided by 1 and itself only with the exception of 1. The fundamental theorem of arithmetic states that any integer can be written as the product of primes and this process is called prime factorization of the integer under consideration. We then described the Sieve of Eratosthenes to list primes up to a given integer and showed an algorithm for the same purpose.

Given two integers a and b, a is said to be congruent to b modulo k if $k \mid (a - b)$ and this relation is shown as $a \equiv b \mod k$. The congruence is an equivalence relation and we reviewed methods to solve linear congruence equations. Lastly, we introduced cryptography which is the science of exchanging messages secretly over a communication media. Two widely used protocols, Diffie-Hellman and RSA are described briefly.

Exercises

Assume a, b, c are integers for all questions.

1. Prove that if $a \mid b$ and $b \mid c$ then $a \mid c$.
2. Prove that if $a \mid b$ and $c \mid d$ then $ac \mid bd$.
3. Prove that if n is odd, $8 \mid n^2 - 1$.
4. Prove that for a positive integer n; if n^2 is a multiple of 3, then n is a multiple of 3.
5. Show that If $a \mid b$ then $\gcd(a, b) = |a|$
6. Prove that If a and b are not both 0, then every common divisor of a and b divides $\gcd(a, b)$.
7. Show the iterations of Euclid's algorithm to find $\gcd(1140, 750)$.
8. Show that $\text{lcm}(na, nb) = |n| \, \text{lcm}(a, b)$.
9. Show that 25 and 33 are coprimes.
10. Find prime factor decompositions of 92, 105 and 213.
11. Find $\gcd(210, 125)$ using prime factorization.
12. Show that Goldbach's conjecture holds for 24, 46 and 78.
13. Write the pseudocode of an algorithm that will perform Eratosthenes method to find primes.
14. Show that for integers a, b, c and d; if $a \equiv b \pmod{m}$ and $c \equiv d \pmod{m}$, then $ac \equiv bd \pmod{m}$.
15. Prove that if $a \equiv b \pmod{m}$, then $a + c \equiv b + c \pmod{m}$ and $ac \equiv bc \pmod{m}$.
16. Find all solutions or conclude no solutions to the linear congruences below:

$$a.\; 5x \equiv 2 \pmod{7} \qquad b.\; 22x \equiv 3 \pmod{44}$$
$$c.\; 3x \equiv 2 \pmod{5} \qquad d.\; 10x \equiv 6 \pmod{14}$$

17. Convert the hexadecimal numbers 2AD, A4C5 and 3A2F to decimal numbers.
18. Convert 123, 618 and 205 to hexadecimal and binary numbers.

References

1. Conradie W, Goranko V (2015) Logic and discrete mathematics. Wiley
2. Goodaire EG, Parmenter MM (2002) Discrete Matheatics with Graph Theory, 2nd Ed., Prentice-Hall
3. Lewis H, Zax R (2019) Essentials discrete mathematics for computer science. Princeton University Press

Counting and Probability

8

Combinatorics is a branch of mathematics that studies the configurations and arrangements of objects. Enumeration is one important task dealt in combinatorics to find
the number of configurations of objects under consideration and counting is the basic
process to perform this task. We need counting when we analyze the complexity of
algorithms. We start this chapter by reviewing basic counting methods and then study
the basic principles of permutations and combinations which are methods to count
the number of ways that a set of objects can be organized. Probabilities of events can
be computed using counting principles as we investigate in the second part of this
chapter.

8.1 Basic Counting Methods

We need to review the principle of inclusion and exclusion before stating two basic
counting principles.

8.1.1 Principle of Inclusion-Exclusion

We have seen the number of elements of a union of sets and intersection of sets in
Chap. 4. The number of elements of A is denoted by $|A|$ and given two sets A and
B, if $C = A \cup B$ and $A \cap B = \varnothing$ then clearly $|C| = |A| + |B|$. However, if the
intersection of the sets A and B is not the empty set, we have the following formula
for the number of elements of the union of these sets,

$$|C| = |A| + |B| - |A \cap B|$$

© Springer Nature Switzerland AG 2021
K. Erciyes, *Discrete Mathematics and Graph Theory*, Undergraduate Topics
in Computer Science, https://doi.org/10.1007/978-3-030-61115-6_8

simply because the common elements of the sets are counted twice. This equation is the simplest form of *inclusion-exclusion principle* which is the determination of the number of elements of the union of two or more finite sets. We will extend these concepts to more than two sets. The number of elements of three sets can be stated as below.

$$|A \cup B \cup C| = |A| + |B| + |C| - |A \cap B| - |A \cap C| - |B \cap C| + |A \cap B \cap C|$$

We add the elements of each set once and need to subtract the number of each pairwise intersections as in three sets. However, elements that are in the intersection of at least three sets are subtracted more than once. In order to correct this situation, we add the number of elements of all possible three set combinations. In both cases, we keep the indices in order such that $i < j$ for two set combinations and $i < j < k$ for three set combinations. We need a more general formula for n sets which can be specified as below.

$$\left| \bigcup A_i \right| = \sum_{i=1}^{n} |A_i| - \sum_{i,j:1 \leq i \leq j \leq n}^{n} |A_i \cap A_j|$$

$$+ \sum_{i,j,k:1 \leq i \leq j \leq k \leq n}^{n} |A_i \cap A_j \cap A_k| + \cdots + (-1)^{n+1}|A_1 \cap \ldots \cap A_n| \tag{8.1}$$

Example 8.1.1 Given sets $A = \{1, 2, 3, 4\}$, $B = \{1, 5, 6, 7\}$, $C = \{1, 3, 5, 6\}$ and $D = \{1, 3, 8, 9\}$, work out the number of elements of the union of these four sets. *Solution*: We use Eq. 8.1 for four sets as below,

$$|A \cup B \cup C \cup D| = |A| + |B| + |C| + |D| - |A \cap B| - |A \cap C| - |A \cap D| - |B \cap C|$$
$$-|B \cap D| - |C \cap D| + |A \cap B \cap C| + |A \cap B \cap D| + |A \cap C \cap D| + |B \cap C \cap D| - |A \cap B \cap C \cap D| \tag{8.2}$$

The number of elements in each set is 4, number of pairwise intersections are 1, 2, 2, 3, 1, 2 in the order presented in Eq. 8.2. The three set intersections are 1, 1, 2, 1 and the four set intersection has 1 element. Note that -1 power is positive for 3-set terms and negative for 2-set and 4-set terms. Substituting these values in Eq. 8.2 yields $4 + 4 + 4 + 4 - 1 - 2 - 2 - 3 - 1 - 2 + 1 + 1 + 2 + 1 - 1 = 9$ which is $|A \cup B \cup C \cup D| = \{1, 2, 3, 4, 5, 6, 7, 8, 9\}|$.   □

Example 8.1.2 A binary number has 0 or 1 as it digits. Find the count of 6-bit binary numbers that start with '11' or end with a '1'.
Solution: Let the set A be the set of 6-bit binary numbers that start with '11' and set B be the 6-bit binary numbers that end with '1'. There are only $2^4 = 16$ 6-bit numbers that start with a '11' ($|A|$). There are only $2^5 = 32$ 6-bit numbers that end with a '1' ($|B|$). Numbers with 6 digits that start with a '11' and end with a '1' have the form '11___1' and there can only be $2^3 = 8$ of such numbers (($|A \cap B|$). Note the use

of word *or* in the question which means we are searching any one of the required conditions to hold meaning union is needed. The required count of such numbers is,

$$|A \cup B| = |A| + |B| + |A \cap B| = 16 + 32 - 8 = 44$$

$\square$

8.1.2 Additive Counting Principle

We will denote the term *event* to mean the outcome of an *experiment*. An experiment can be tossing a coin, drawing a card etc. Let A be an event corresponding to an experiment. Then $|A|$ is the number of possible outputs of A, for example, if A is flipping a coin, then $|A|$ is 2 since we can only have heads or tails. Two events A and B are mutually exclusive if they cannot occur together. We are now ready to define the additive principle as below.

Definition 8.1 (*additive counting principle*) Assume some event E can occur in n possible ways and a second mutually exclusive event F can occur in m possible ways, and assume both events cannot occur simultaneously. Then E *or* F can occur in $n + m$ different ways.

Example 8.1.3 Consider three cities A, B and C. One can take ferry, bus, train or drive by car from city A to city B. It is possible to travel from A to C by car, train or bus. In how many ways one can travel from city A to city B or city C?
Solution: Let event E_1 be traveling from A to C and E_2 be traveling from B to C. Possible ways to perform E_1 is 4 and E_2 is 3. Thus, total number of ways of traveling from A to B, or from A to C is $4 + 3 = 7$. $\square$

Example 8.1.4 How many two-letter words start with letter A or B in English?
Solution: Let event E_1 be two-letter words starting with A and E_2 be two-letter words starting with B. There are 26 letters in the English alphabet, so we can have 26 words starting with A and 26 words starting with letter B. Therefore $|E_1| = 26$ and $|E_2| = 26$. Thus, total number of words that start with either A or B is $26 + 26 = 52$. $\square$

8.1.3 Multiplicative Counting Principle

The multiplicative counting principle (MCP) can be stated as follows.

Definition 8.2 Assume some event E can occur in n possible ways and a second event F can occur in m possible ways, and assume both events cannot occur simultaneously. Then E and F can occur in sequence in $n \times m$ different ways.

In general, there may be m events $E_1, E_2, \ldots, E_m$ with E_1 occurring in n_1 different ways, E_2 in n_2 different ways and so on with E_m having n_m different ways of executions. We could then have $n_1 \times n_2 \times \cdots \times n_m$ different executions of events in total. Note that occurring of event E_k is based on assuming all events $E_1, \ldots, E_{k-1}$ have already occurred. For example, assume a student wants to take a math course out of 3 possible courses M_1, M_2 and M_3 and a programming course out of 4 courses, P_1, P_2, P_3 and P_4. Assuming she has taken M_1 (event E_1 has occurred), she can take one of four programming courses afterwards. For each math course, there are 4 possible proceeding programming courses. Thus, the number of possible course allocation events is $3 \times 4 = 12$ for this student.

Example 8.1.5 How many positive integers of 4 digits can be generated which do not contain the same digit twice? How many 4-digit numbers can be generated that end with an even digit allowing repeated digit values?
Solution: For the first part of the question; the first digit may contain 9 numbers (1–9), the second digit may again contain 9 numbers, this time including zero but excluding the first digit which is now fixed. The third digit may contain 8 and the last digit may have any of the remaining 7 numbers. Total count of integers is,

$$9 \times 9 \times 8 \times 7 = 4536.$$

For the second part; the first digit may have 9 numbers, the second and third digits may have 10 possibilities and the last digit may only be 0, 2, 4, 6 and 8 for a total of 5 possibilities. Thus, total count of such numbers is,

$$9 \times 10 \times 10 \times 5 = 4500.$$

□

Example 8.1.6 Let us work out the possible numbers that contain odd digits that can be output by a 3-digit display. Note that we do not just require an odd number but an integer that contains *all* odd digits. Each digit may have 1, 3, 5, 7 or 9 for a total of 5 odd digits. Let displays of hundreds digit be E_1, tens digit be E_2 and unit digit be E_3 events. Total number of possible events is $5 \times 5 \times 5 = 125$. If repetition of digit values were not allowed, we would have $5 \times 4 \times 3 = 60$ possible events. □

Example 8.1.7 How many numbers in the range 100–999 do not have repeated digits?
Solution: We can have 9 digits between and including 1 and 9 for the first digit excluding zero. The second digit may contain a zero and we have now 9 digits for the second digit excluding the first assigned digit. The third digit will have 8 digits out of 10 excluding the two digits already assigned. The count of numbers is then $9 \times 9 \times 8 = 648$. Note that we assume the first event (assignment of the fist digit) has occurred when searching the possibilities of the second digit assignment. □

A computer program may have nested loops as shown in Alg. 8.1. The most inner loop with index k will run to completion executing p times for each value of the middle loop with index j. Similarly, the middle loop will run for m times for each value of the outermost loop. Thus, total number of executions of such a program is the product of the number of times each loop executes which is nmp for this example.

Algorithm 8.1 *Nested Loops*

1: **for** $i = 1$ to n **do**
2:     **for** $j = 1$ to m **do**
3:         **for** $k = 1$ to p **do**
4:             *print i, j, k*
5:         **end for**
6:     **end for**
7: **end for**

8.1.4   The Pigeonhole Principle

Let us assume there are m pigeons to be placed in n pigeonholes with $m > n$. When the pigeons are placed in the pigeonholes, some pigeonholes will have more than one pigeon in it.

Example 8.1.8 Prove the pigeonhole principle using the contrapositive method.
    The contrapositive of this principle is that if every pigeonhole contains at most one pigeon, then $m \leq n$. Let a_i be the number of pigeons in pigeonhole i, we have;

$$m = \sum_{i=1}^{n} a_i$$

as m is the number of total pigeons.

$$m = \sum_{i=1}^{n} a_i \leq \sum_{i=1}^{n} 1 = n$$

Thus, $m \geq n$.   □

Example 8.1.9 There are 13 workers in an office. Then, at least two of the workers were born in the same month.   □

Remark 8.1 (*general pigeonhole principle*) Assume there are $kn + 1$ pigeons to be assigned to n pigeonholes with $k \in \mathbb{N}$. Then, there is at least one pigeonhole assigned to $k + 1$ or more pigeons.

Example 8.1.10 Find the minimum number of a group of friends such that four are born in the same day of the week.

Solution: There are 7 days of the week, hence $n = 7$. Also, $k + 1 = 4$, therefore $k = 3$. We can conclude that $kn + 1 = 22$ which means there should be at least 22 friends in the group to ensure at least four are born in the same day of the week. □

8.1.5   Permutations

Let us assume we have n objects to be arranged. Our aim is to find the number of all possible ordered arrangements of these objects, for example, let $S = \{a, b, c\}$. The possible ordered arrangements of these objects are abc, acb, bac, bca, cab and cba for a total of 6 ways. For a set containing n objects, here are n objects to be placed in the first position, so there are n ways, the second position may hold $n - 1$ objects and so on, and by the multiplicative principle, we can have $n \times (n - 1) \times \cdots \times 1 = n!$ (n factorial) ordered arrangements in total.

Definition 8.3 (*permutation*) A permutation of n distinct objects is an *ordered arrangement* of these objects and is equal to $n!$.

Example 8.1.11 Find the number of ways that the letters in the word ALGORITHM be arranged. If the three letters ALG should be next to each other, work out the number of ways any word with this property can be written using these letters.
Solution: All the letters in this word are distinct. Let the first letter to be placed be A, it can be placed in one of the 9 positions. Once it has been placed, the next letter can be in one of the remaining 8 places. Thus, following in this manner, the number of possible different writings of any word using these letters is $9! = 362,880$. For the second part of the question, the word ALG is treated as one unit and we then have 7 places instead of 9. Therefore, we will have $7! = 5,040$ different words with this property. For example, RMALGIOHT is one such word. □

In the more general case, we may have a certain number of objects to be selected from a set of n objects. For example, let us find the possible ordered arrangements of size 3 from the elements of the set $\{a, b, c, d\}$. These arrangements are,

$$abc, abd, acb, acd, adc, adb, bac, bad, bca, bcd, bda, bdc$$

$$cab, cad, cba, cbd, cda, cdb, dab, dac, dba, dbc, dca, dcb$$

for a total of 24 items. Displaying all possible orderings and counting them will be difficult when the size of the objects is large, thus, we need to formalize the number of this arrangements.

Definition 8.4 (*r-permutation*) An r-permutation of a set of n objects is an ordered subset of r elements selected from n objects.

Theorem 13 *The number of r-permutations of a set of n objects where* $n \in \mathbb{Z}^+$ *and* $0 \le r \le n$ *is,*

$$P(n, r) = \frac{n!}{(n - r)!} \tag{8.3}$$

Proof There are n ways to select the first element and once it is selected, there are $n - 1$ was to select the second one until there are $n - r + 1$ ways to select the rth element. This is the product $n \times (n - 1) \times \cdots \times (n - r + 1)$ which can be stated as,

$$P(n, r) = \frac{n!}{(n - r)!}$$

□

Example 8.1.12 Find the number of ways that the letters ABCDE can be arranged as two letters.

Solution: We apply Eq. 8.3 to yield the following,

$$P(5, 2) = \frac{5!}{(5 - 2)!} = 60$$

□

Example 8.1.13 Consider the following:

1. How many ways 4 different colors (blue (B), green (G), white (W), yellow (Y)) of marbles be ordered in groups of 3?
2. If the first marble is green, what is the number of possible arrangements and what are they?

Solution: The number of colors is 4 which is n, and we need subsets of r which is 3. Thus,

$$P(4, 3) = \frac{4!}{(4 - 3)!} = 4$$

For the second part, when the first marble is fixed as yellow, we have only two places left for the remaining 3 colors. The $(3, 2)$ permutations is then $3!/(3-2)! = 6$. These are YBG, YBW, YGB, YGW, YWG and YWB.

□

Theorem 14 (permutations with repetitions) *Consider a set with n elements such that* n_1 *elements are alike,* n_2 *elements are alike, ...,* n_r *elements are alike. Then n permutations of this set is,*

$$P(n; n_1, n_2, \ldots, n_r) = \frac{n!}{n_1! \cdot n_2! \cdots n_r!}$$

Proof We will illustrate the proof by an example. Let us apply subscripts to objects of the same type. For example, the word $AMANDA$ becomes $A_1MA_2NDA_3$. The number of possible objects are reduced from 6 to 4 (A, M, N, D). The order of A's is not important when the order of subscripted objects is changed. For example, $A_1A_2A_3MND$ or $A_3A_1A_2MND$ are the same and $A_2NDA_1MA_3 = A_1NDA_3MA_2$. Therefore, we need to divide the possible number of permutations with the permutations of the repeated objects. In this example, we can have $6!/3! = 120$ different words. When there are other repeated objects, the result can be generalized.   □

Example 8.1.14 Let us find the number of ways to write different words from the word $ALIBABA$. Closer look reveals A is repeated three times and B is repeated twice. Therefore,

$$P(7; 2!, 3!) = \frac{7!}{2!3!} = 420$$

□

Remark 8.2 (*circular permutation*) Assume there are n objects to be arranged in a circle. Then, these objects can be arranged in $(n - 1)!$ ways.

Example 8.1.15 The number of ways 8 people can sit around a table is $(8 - 1)! = 7! = 5040$.   □

8.1.6 Combinations

Let us consider the set $A = \{a, b, c, d\}$ again and we want to list all possible subsets of A that have 3 elements with the property that no subset should be equal to another one. The order of the elements in selections from the set A is not important, for example, we can take only $\{a, b, c\}$ or $\{b, a, c\}$ but not both. The possible 3 element selections are,

$$abc, abd, bcd, dac$$

Such selections from a set are called *combinations*.

Definition 8.5 (*combination*) An r-combination ($C(n, r)$) or $\binom{C}{r}$) of n distinct objects of a set where $n \in \mathbb{Z}^+$ and $0 \leq r \leq n$ is the subsets of the set without regarding the order of elements in the subset.

Theorem 15 *The number of r-permutations of a set of n objects is,*

$$\binom{n}{r} = C(n, r) = \frac{n!}{r!(n - r)!} \tag{8.4}$$

Proof Given n objects, the number of r-permutations is $P(n, r)$. Each of these permutations can be ordered in $P(r, r)$ ways which is $r!$. The total number of r-combinations is therefore the ratio of total number of permutations to the number of ways each permutation can be ordered as follows,

$$C(n, r) = \frac{P(n, r)}{P(r, r)} = \frac{n!/(n - r)!}{r!/(r - r)!} = \frac{n!}{r!(n - r)!}$$

□

In our first example with $A = \{a, b, c, d\}$, $n = 4$ and $r = 3$ yields,

$$C(4, 2) = \frac{4!}{3!(4 - 3)!} = 4$$

as we have found.

Example 8.1.16 The number of 3-combinations of digits $0, 1, \ldots, 6$ is;

$$\frac{7!}{3!4!} = \frac{7 \cdot 6 \cdot 5 \cdot 4 \cdot 3 \cdot 2 \cdot 1}{3 \cdot 2 \cdot 1 \cdot 4 \cdot 3 \cdot 2 \cdot 1} = 35$$

with 521 being such one combination.   □

Example 8.1.17 A group of three men and four women are to be chosen from a group of five men and six women. How many possibilities are there?
Solution: Three men can be chosen in $C(5, 3)$, and four women can be selected in $C(6, 4)$ ways. Using multiplicative counting principle, total number of selections is,

$$C(5, 3) \times C(6, 4) = \frac{5!}{3!2!} \times \frac{6!}{4!2!} = 10 \times 15 = 150$$

□

Theorem 16 *For all integers $n \geq 1$ and all integers r, $1 \leq r \leq n$,*

$$\binom{n}{r} = \binom{n}{n - r}$$

Proof

$$\binom{n}{r} = \frac{n!}{r!(n - r)!} = \frac{n!}{(n - r)!(n - (n - r)!)} = \binom{n}{n - r}$$

□

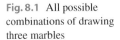

Fig. 8.1 All possible
combinations of drawing
three marbles

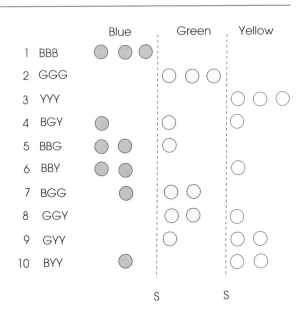

8.1.6.1 Combinations with Repetitions

Consider an experiment in which we have a box containing blue (B), green (G) and yellow (Y) marbles and we select 3 marbles at random. How many different combinations are possible? The possible combinations may be listed as follows.

BBB, GGG, YYY, BGY, BBG, BBY, BGG, GGY, GYY, BYY,

for a total of 10 combinations. We can not use the usual combination formula this time. Let us consider each marble color as a distinct class separated by a division as shown in Fig. 8.1.

We now have either a class of marbles or two separators to choose from, for a total of 5 positions; 3 for the marble colors and 2 for the separators. Our each selection may be a combination of colors and separators (S), for example BBSGS is BBG (line 5 in figure) or BSSYY is BYY (last line in figure) discarding the separators. As a result, we select 3 positions out of 5, thus, $C(5, 3) = 10$ as found before. A generalization of this result follows. Consider a set A that contains k different kinds of elements with at least m elements of each kind. Then, the number of possible combinations when m items are selected is,

$$\binom{m + k - 1}{m} \tag{8.5}$$

We need to select m elements and $k - 1$ is the number of separators between them, thus, the total number of selections we can make is $m + k - 1$ and we need to have m selections. When we apply this formula to the above example, we have 3 ($k = 3$) types of objects (B,G,Y) and we should have at least 3 ($m = 3$) of each to be able to select BBB for example,

$$\binom{3+3-1}{3} = C(5,3) = 10$$

Example 8.1.18 Consider a jar containing sufficient amount of marbles of four colors; blue (B), green (G), yellow (Y) and white. How many combinations are possible if three marbles are drawn from the jar?

Solution: We have 4 colors (B, G, Y, W) and make 3 selections. Thus, $k = 4$ and $m = 3$, substitution in Eq. 8.5 yields,

$$\binom{4+3-1}{3} = C(6,3) = 20$$

These possibilities are as follows,

BBB, GGG, YYY, WWW, BGG, BYY, BWW, GBB, GYY, GWW
YBB, YGG, YWW, WBB, WGG, WYY, BGY, BGW, BYW, GYW

8.1.6.2   The Binomial Theorem

The following equations can be proven without difficulty.

$$(x + y)^0 = 1$$
$$(x + y)^1 = x + y$$
$$(x + y)^2 = x^2 + 2xy + y^2$$
$$(x + y)^3 = x^3 + 3x^2y + 3xy^2 + y^3$$

The *Binomial Theorem* provides a generalization for the expansion of $(x + y)^n$ as follows.

$$(x + y)^n = \sum_{k=0}^{n} \binom{n}{k} x^{n-k} y^k$$

$$= \binom{n}{0} x^n + \binom{n}{1} x^{n-1} y + \binom{n}{2} x^{n-2} y^2 + \cdots + \binom{n}{n-1} xy^{n-1} + \binom{n}{1} x^{n-1} y + \binom{n}{n} y^n$$

Example 8.1.19 Find expansion of $(x + y)^5$ using Binomial theorem.
Solution:

$$(x + y)^5 = \sum_{k=0}^{5} \binom{5}{k} x^{5-k} y^k$$

$$= \binom{5}{0} x^5 + \binom{5}{1} x^4 y + \binom{5}{2} x^3 y^2 + \binom{5}{3} x^2 y^3 + \binom{5}{4} xy^4 + \binom{5}{5} y^5$$

$$= x^5 + 5x^4 y + 10x^3 y^2 + 10x^2 y^3 + 5xy^4 + y^5$$

$\square$

It can be shown that,

$$\binom{n}{k} = \binom{n}{n-k} \tag{8.6}$$

and

$$\binom{n+1}{k+1} = \binom{n}{k} = \binom{n}{k+1} \tag{8.7}$$

using the definition of permutation (See Example 10).

Pascal's Triangle

The binomial coefficients can be computed simply and efficiently using the following arrangement.

$$
\begin{array}{ccccccccccc}
& & & & & \binom{0}{0} & & & & & \\
& & & & \binom{1}{0} & & \binom{1}{1} & & & & \\
& & & \binom{2}{0} & & \binom{2}{1} & & \binom{2}{2} & & & \\
& & \binom{3}{0} & & \binom{3}{1} & & \binom{3}{2} & & \binom{3}{3} & & \\
& \binom{4}{0} & & \binom{4}{1} & & \binom{4}{2} & & \binom{4}{3} & & \binom{4}{4} & \\
\binom{5}{0} & & \binom{5}{1} & & \binom{5}{2} & & \binom{5}{3} & & \binom{5}{4} & & \binom{5}{5}
\end{array}
$$

....

 This triangle has the property that each row starts with a 1 and ends with a 1. Moreover, any number other than the 1s in the triangle is the sum of the two numbers that appear at its upper left and upper right corners by Eq. 8.7. The following arrangement called *Pascal's Triangle* implements these observations and provides a fast way to compute binomial coefficients.

$$
\begin{array}{ccccccccccccc}
& & & & & & 1 & & & & & & \\
& & & & & 1 & & 1 & & & & & \\
& & & & 1 & & 2 & & 1 & & & & \\
& & & 1 & & 3 & & 3 & & 1 & & & \\
& & 1 & & 4 & & 6 & & 4 & & 1 & & \\
& 1 & & 5 & & 10 & & 10 & & 5 & & & \\
1 & & 6 & & 15 & & 20 & & 15 & & 6 & & 1
\end{array}
$$

....

8.2   Discrete Probability

Definition 8.6 (*sample space, event*) The set S of all possible occurrences of an experiment is called the *sample space* of the experiment. An *event* is a subset of the sample space S.

For example, when tossing of a coin is the experiment, the sample space S is {heads, tails}. An event of this experiment is either *heads* (H) or *tails* (T) but not both. It is of practical use to think an event as a subset of the sample space to be able to implement set properties on the event. The following properties of events are immediately available when two events A and B are considered.

- $C = A \cup B$ is the event that occurs when A or B or both occur.
- $C = A \cap B$ is the event that occurs when A and B or both occur.
- A^c (or $\overline{A}$) is the event that occurs when A does not occur.

Example 8.2.1 What is the number of possibilities when a pair of dice of different colors is thrown?
Solution: Each dice can have one of {1, 2, 3, 4, 5, 6}. The sample space therefore consists of {(1, 1), (1, 2), ..., (1, 6), (2, 1), ..., (2, 6),.....(6, 6)} 36 pairs using the multiplicative counting principle. □

Example 8.2.2 What is the number of possibilities when a coin is tossed and a dice is thrown? What is the number of possible events to have H and an odd number outcome? □
Solution: The sample space of this experiment is {$(H,1), (H,2), (H,3), (H,4), (H,5)$, $(H,6), (T,1), (T,2), (T,3), (T,4), (T,5), (T,6)$} for a total of 12 possible outcomes. For the second part of the question, we have {$(H,1), (H,3), (H,5)$} as the three possible outcomes.

8.2.1   Probability Measures

Definition 8.7 (*probability, probability distribution*) The numerical measure of the likelihood of an event is called the *probability* ($P(E)$) of an event E. A set of the probabilities of all outcomes of an experiment is called the *probability distribution* of the experiment.

Let us consider the experiment of flipping a coin. The following probabilities of heads (H) and tails (T) can be stated.

$$P(H) = P(T) = \frac{1}{2}.$$

When each possible event of an experiment has equal chance, the probabilities of these events are equal as in the case of tossing a coin. The probability distribution of the experiment is then termed *uniform* and the occurrence of each event has a probability of $1/n$ when n such events are considered.

Definition 8.8 (*classical probability*) Consider an experiment with a finite number of uniform events. Then the *classical probability* of an event is defined to be the ratio of the number of possible expected (favorable) outcomes to the total number of possible outcomes of the experiment.

Example 8.2.3 What is the probability to have an odd number outcome when a dice is thrown?
Solution: The sample space of this experiment is $\{1, 2, 3, 4, 5, 6\}$ with the size 6. The favorable outcomes are 1, 3 and 5 for a total of three numbers. Thus, the probability of receiving an odd number is $3/6 = 0.5$.   □

Example 8.2.4 A deck of cards is shuffled and a card is drawn. What is the probability of drawing this card as hearts?
Solution: There are 13 hearts in a deck of cards with 52 cards. The probability of drawing a hearts from the deck of cards is then,

$$P(E) = \frac{|E|}{|S|} = \frac{13}{52} = 0.25$$

□

Example 8.2.5 A pair of dice with distinct colors is thrown. What is the probability of having a throw such that the sum of the numbers on the top of die is 5?
Solution: There are 6 numbers, 1,…,6, on the faces of a dice. When two die are thrown, there are 36 possible pairs of numbers on top by the multiplicative counting principle. Out of these, (1, 4), (2, 3), (3, 2), (4, 1) provide the required result. The probability of the favorable events is then,

$$P(E) = \frac{|E|}{|S|} = \frac{4}{36} = 0.1\overline{1}$$

□

Theorem 17 *Let S be a sample space and E, E_1 and E_2 be any event of this sample space. Then, the following can be stated:*

- $0 \leq P(E) \leq 1$
- $P(S) = 1$.
- *If two events E_1 and E_2 of S are mutually exclusive; that is, they cannot both happen, $P(E_1 \cup E_2) = P(E_1) + P(E_2)$*
- *Let E^c be the event that event E does not occur. Then $P(E^c) = 1 - P(E)$.*
- *In the more general case, $P(E_1 \cup E_2) = P(E_1) + P(E_2) - P(E_1 \cap E_2)$*
- *$P(E_1 \setminus E_2) = P(E_1) - P(E_1 \cap E_2)$*

Example 8.2.6 A deck of cards is shuffled and a card is drawn. What is the probability of drawing this card as spades or an odd number?

Solution: There are 52 cards in a deck 13 out of which is spades. The odd numbered cards have 3, 5, 7 and 9 on them for a total of 16 for four different symbols. Let A be the event that the drawn card is spades and B the event that it has an odd number. The number of spade cards that have odd numbers are 4. Then,

$$P(A \cup B) = P(A) + P(B) - P(A \cap B) = \frac{13}{52} + \frac{16}{52} - \frac{4}{52} = \frac{33}{52} = 0.65$$

□

Example 8.2.7 What is the probability of getting a double or a sum of odd number when throwing two die?

Solution: Let event A be "getting a double" and B be "sum is an odd number". These two events are mutually exclusive as all of the sums of doubles are even numbers. Note that the sum of an odd number by itself and the sum of an even number by itself both yield even numbers. The event A happens in 6 throws out of 36, and B happens in 12 events out of 36. Then,

$$P(A \cup B) = P(A) + P(B) = \frac{1}{6} + \frac{2}{6} = 0.5$$

□

8.2.2   Independent Events

Definition 8.9 (*independent events*) Two events A and B are independent if $P(A \cap B) = P(A) \cdot P(B)$, otherwise they are dependent.

Example 8.2.8 Let us find the probability of having two heads on independent coin rolls. Let event A be having a head in the first roll and B be having a head in the second roll, and since $P(A) = P(B) = 1/2$, $P(A \cap B) = 1/2 \cdot 1/2 = 1/4$.   □

Example 8.2.9 A coin is rolled twice. What is the probability of having a Heads (H) and Tails (T) consecutively?

Solution: Let HT denote the required sequence of events. Since each coin roll is independent of the other and we have $P(H) = P(T) = 1/2$,

$$P(A \cap B) = P(A) \cdot P(B) = \frac{1}{2} \cdot \frac{1}{2} = \frac{1}{4}$$

Note that any coin rolling combination in two rolls will have the same probability. We can check whether events A and B are independent by testing $P(A \cap B) = P(A) \cdot P(B)$. Indeed, HT roll has the probability of 1/4 out of 4 possible two rolls of HH, HT, TH and TT.   □

Example 8.2.10 A pair of distinct dice is rolled. Let event A be "the sum of the faces is 8" and event B be "the first die is 2". Test whether A and B are independent.
Solution: Let us check using the definition of independent events. The event $P(A \cap B) = 1/36$ (throw $(2,6)$) as this event is a single outcome out of 36 events. $P(A) = P(B) = 1/6$ and thus,

$$P(A \cap B) = \frac{1}{36} = \frac{1}{6} \cdot \frac{1}{6} = P(A) \cdot P(B)$$

Therefore, these events are independent.

8.2.3 Conditional Probability

Let two events B and A be the output of two consecutive experiments on a sample space S. It is possible that the probability of event A is influenced by the output of event B. In this case, conditional probability of event A given event B is defined as follows.

Definition 8.10 (*conditional probability*) The probability of an event A once an event B has occurred is called the *conditional probability* of A given B, provided $P(B) > 0$, is defined as follows.

$$P(A \mid B) = \frac{P(A \cap B)}{P(B)} \tag{8.8}$$

Example 8.2.11 A group of 40 students need to elect courses. Math is elected by 20, Physics is elected by 10 and 5 of the students elect both Math and Physics. If a student has elected Math at random, what is the probability that this student also elected Physics?
Solution: Let event A be selection of Physics, and B be selection of Math. Then,

$$P(A \mid B) = \frac{P(A \cap B)}{P(B)} = \frac{5/40}{20/40} = 0.25$$

$\square$

Example 8.2.12 A 3-bit binary number is generated at random. What is the probability that the generated number has two consecutive 1s given the last digit is a 1?
Solution: Let event B be the generation of a binary number with a 1 in the last digit and A be he event that a generated number has two consecutive 1s. There are eight possible 3-bit numbers; $000, 001, 010, 011, 100, 101, 110, 111$. Only 4 of the generated numbers end with a 1 ($001, 011, 101, 111$) and three of the generated numbers ($011, 110, 111$) have two consecutive 1s. The intersection of these two sets is ($011, 111$). Thus,

$$P(A \mid B) = \frac{P(A \cap B)}{P(B)} = \frac{2}{4} = 0.5$$

It can be seen that 011 and 111 provide the required condition once the last digit is a 1. Since there are 4 numbers with last digit 1, the conditional probability is 0.5. $\square$

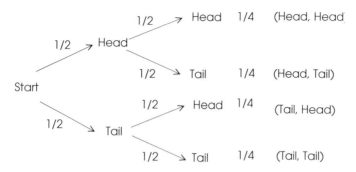

Fig. 8.2 Tree diagram for rolling a coin twice

Corollary 18 *Let A and B two events with $P(B) > 0$. Then,*

$$P(A \cap B) = P(A \mid B) \cdot P(B)$$

Example 8.2.13 Let event B be "getting flu" and event A be "having a high fever" with $P(B) = 0.2$, $P(A \mid B) = 0.8$ (having high fever when flu is present). What is the probability of getting flu and fever?
Solution: We can use Corollary 18 as follows.

$$P(A \cap B) = P(A \mid B) \cdot P(B) = 0.8 \cdot 0.2 = 0.16$$

□

8.2.4 Tree Diagrams

A *tree diagram* is a visual method of showing the resulting choices and probabilities of two or more events. A rooted tree basically consists of a starting node called the *root*, nodes and branches which are edges connecting nodes. The end of each branch of a tree is labeled by the outcome of the events starting from the root of tree and ending at the end of the branch. Each branch is labeled with the probability of the event that starts from one end of the branch to its other end. Let us consider an experiment of rolling a coin two times. The tree diagram for this experiment is depicted in Fig. 8.2. We start from an initial state and the chances of rolling a Head or Tail is 1/2 each as shown by the labels of branches. The probability of the final leaves of the tree are calculated by multiplying the probabilities of each branch leading to the leaf. For example, the bottom leaf is rolling Tail and Tail resulting in $(1/2) \cdot (1/2) = 1/4$.

Example 8.2.14 A jar contains 5 blue (B) and 3 green (G) marbles. Work out the probabilities of picking a marble three times consecutively from this jar without putting them back by using a tree diagram.
Solution: Note this time we have dependent events as picking a certain color affects

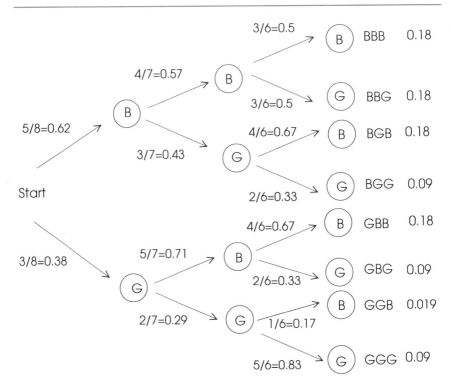

Fig. 8.3 Tree diagram for Example 8.2.14

the consequent probabilities and we need to calculate conditional probabilities. The
tree diagram has a level of three, starting from the root as shown in Fig. 8.3. The sum
of the probabilities of branches coming out of a node must equal 1 since this sum is
the probabilities of all events.

□

8.2.5 Random Variables

Definition 8.11 (*random variable*) Let S be the sample space of an experiment. A
random variable X is defined as a function from S to the real numbers: $X : S \rightarrow \mathbb{R}$.

In other words, a random variable assigns a real number to every possible outcome
of a random experiment. The set of values that a random variable X can take is called
the *range* of X.

Example 8.2.15 Let the random experiment be rolling a coin twice. The sample
space S is $\{H, T\}$, and let a random variable X defined as the number of heads.
Then, $X(HH) = 2, X(HT) = 1, X(TH) = 1, X(TT) = 0$.

$$X : S \to \mathbb{R}$$

☐

A random variable X is is a *discrete random variable* if there is a finite number of possible outcomes of X, in other words, its range consists of finite or countable values or there is a countably infinite number of possible outcomes of X.

Example 8.2.16 Let us consider the experiment of rolling a a pair of distinct dice, one blue and one green and let the random variable X denote the sum of the faces of each die. Then, X has a range $1,...,12$.   ☐

Definition 8.12 (*independent random variable*) Two random variables X and Y defined over a sample space S are independent if,

$$P(X = x \cap Y = y) = P(X = x) \cdot P(Y = y)$$

Note that this definition is similar to what we have for two independent events. For example, let random variable X be the number on the face of the first die and random variable Y be the number on the face of the second die when rolled. Then X and Y are independent.

8.2.5.1 Expectation of a Random Variable
The mean, expected value or expectation of a random variable X, $(E(X))$, is the long term average of the random variable.

Definition 8.13 (*expectation*) Expectation $E(X)$ of a real-valued random variable X is the average value of X weighted by probability. For discrete random variables, this parameter is defined by the following where the probability that X maps into x is denoted by $P(X = x)$.

$$E(X) = \sum_{x \in S} x \cdot P(X = x)$$

Example 8.2.17 Consider the experiment of rolling a dice. The sample space is $1, 2, 3, 4, 5, 6$. Let the random variable X be the number on the face of the dice when rolled and find the expectation of X.

Solution: X can have values in the range $1,...,6$. The expectation $E(X)$ is then,

$$1 \cdot (1/6) + 2 \cdot (1/6) + 3 \cdot (1/6) + 4 \cdot (1/6) + 5 \cdot (1/6) + 6 \cdot (1/6) = 7/2$$

☐

Example 8.2.18 Let us consider the experiment of rolling a coin twice again and work out $E(X)$ with X being the number of heads.
Solution: $P(X)$ for any $x \in S$ is 1/4 since we have 4 possible outcomes. Then,

$$E(X) = \sum_{x \in S} x \cdot P(X = x)$$

$$= \frac{1}{4}(HH) + \frac{1}{4}(HT) + \frac{1}{4}(TH) + \frac{1}{4}(TT)$$

$$= \frac{1}{4}(2 + 1 + 1 + 0)$$

$$= 1.$$

□

We may need to asses the distance of a random variable from its expectation.

Definition 8.14 (*variance*) Let X be a random variable defined on a sample space S. The variance of X denoted by $V(X)$ is defined by,

$$V(X) = E((X - E(X))^2)$$

The standard deviation σ of X is defined by,

$$\sigma(X) = \sqrt{V(X)}$$

These two parameters of a random variable help to assess the divergence of its value from its expectation.

8.2.6 Stochastic Processes

In some cases, an experiment can be divided into a number of sequential processes. Each such process may be a random variable and a stochastic process which is a collection of these random processes may represent a system, typically randomly changing with time. For example, the movement of a gas molecule, photon emission, occurrences of earthquakes and the growth of bacteria population may be modeled as stochastic processes. Informally, a stochastic process is a process that develops in time according to probabilistic rules.

Definition 8.15 (*stochastic process*) A *stochastic process* is a family of random variables $\{X(t), t \in T\}$, where t usually denotes time. In other words, a random variable is assigned to every t value in the set T.

$\{X(t), t \in T\}$ is *discrete-time process* if T is finite or countable, otherwise it is defined as a *continuous-time process*. Stochastic processes may be used in the areas of economics for stock exchange; in epidemiology to monitor the number of influenza cases for example and in medicine to test the effects of a drug.

8.3 Review Questions

1. State the principle of inclusion and exclusion.
2. What is additive counting principle? Give an example.
3. What is multiplicative counting principle? Give an example.
4. Give an example of pigeonhole principle.
5. Define the permutation of n distinct objects.
6. How does repeated permutation differ from permutation of distinct objects?
7. Describe the combination of r objects from n distinct objects.
8. Compare repeated combination with repeated permutation.
9. State Binomial theorem.
10. What is an event and sample space of an experiment?
11. Define the probability of an event.
12. Give an example of two independent events. What is the probability of two independent events occurring?
13. Define conditional probability and state the formula for the conditional probability of event A given that event B has occurred.
14. What does a tree diagram provide?
15. What is a random variable and what is the expectation of a random variable?
16. Describe a stochastic process and give an example.

8.4 Chapter Notes

We reviewed basic counting methods and probability in this chapter. Two main methods of counting are the additive and the multiplicative counting principles. The additive counting principle states that the number of possible ways that two independent events can happen is the sum of the number of ways each can happen, and the multiplicative counting principle means that the number of possible ways that two events that do not occur at the same time is the product of the number of ways each can happen. The pigeonhole principle asserts if there are m objects to be placed in n places with $m > n$, then some places will have more than one object.

A permutation of n distinct objects is an ordering of these objects; some of the objects may be equal to each other in which case we need to consider permutations with repetitions. A combination of a set of n distinct objects is a selection of a subset of these objects. Unlike permutation, we can have only one selection with n elements as the combination of n distinct objects. When some elements of the set of objects are the same, we have combinations with repetitions. The binomial theorem specifies the expansion of the algebraic expression of the form $(a + b)^n$.

In the second part of this chapter, we reviewed basic principles of discrete probability which is the likelihood of the occurrence of an event. The events under consideration may be independent or may depend on each other and conditional probability is used to find probabilities of dependent events. A random variable is a real number assigned to the outcome of an experiment. Expectation of a random variable is the average value of a random variable weighted by its probability. A stochastic process is a sequence of processes output of each depends on the preceding process. The brief review of probability presented here may serve as a starting point for further study of this fundamental mathematics topic.

Exercises

1. In a standard deck of 52 playing cards, in how many different ways can four-of-a-kind (same symbol with hearts, diamonds, spades or clubs) be formed?
2. Consider selecting 2 bananas, 3 apples and 2 kiwis from a box of 6 bananas, 5 apples and 4 kiwis. How many choices can be made?
3. Work out the minimum number of students in a meeting to be sure that three of them are born in the same month.
4. Find the number of ways that 8 people can sit in a row of chairs and around a circular table.
5. Determine the number of 5-element subsets of the set $A = \{1, 2, \ldots, 8\}$.
6. Find the number of n distinct words that can be formed from the word MISSISIPPI.
7. How many subsets of set $A = \{1, 2, \ldots, 9\}$ has three or more elements?
8. A company has 12 workers with 8 men and 4 women. Find the number of ways to select a 5-member committee from these workers. Also, work out the number of ways to select a 5-member committee with 3 men and 2 women.
9. If $\binom{n}{3} = 20$, then what is $\binom{n}{5}$?
10. Prove the following equivalences.

a.
$$\binom{n}{r} = \binom{n}{n-r}$$

b.
$$\binom{n+1}{r+1} = \binom{n}{r} = \binom{n}{r+1}$$

11. Two fair dice are rolled. What is the probability of getting 7 as the sum of two face values?
12. Consider the experiment of rolling a coin three times and work out $E(X)$ with X being the number of heads.

Boolean Algebras and Combinational Circuits

9

Boolean algebra operating on the binary numbers 0 and 1 was first developed by George Boolean in 19th century [1]. A Boolean function is an expression using binary variables. We review basic Boolean algebra laws, duals of these laws, functions and a visual method called Karnaugh-maps to simplify a Boolean expression in the first part of this chapter. We then review combinational circuits in the second part and describe simple structures called logic gates to build a logic circuit to represent a Booelan function. We conclude this chapter by arithmetic circuits to add binary numbers.

9.1 Boolean Algebras

Definition 9.1 (*boolean algebra*) A boolean algebra B consists of a set S with elements 0 and 1 and binary operations $+$ and $\cdot$ on S and a unary operator $'$ on S obeying the following laws for binary variables a, b and c. A Boolean algebra B is denoted by $B = (S, +, \cdot, ', 0, 1)$.

- *Identity Laws*:
$$a + 0 = a, \qquad a \cdot 1 = a$$

- *Idempotent Laws*:
$$a + a = a, \qquad a \cdot a = a$$

- *Complement Laws*:
$$a + a' = 1, \qquad a \cdot a' = 0$$

- *Commutative Laws*:
$$a + b = b + a, \qquad a \cdot b = b \cdot a$$

© Springer Nature Switzerland AG 2021
K. Erciyes, *Discrete Mathematics and Graph Theory*, Undergraduate Topics in Computer Science, https://doi.org/10.1007/978-3-030-61115-6_9

Table 9.1 Representations of logical variables

Description	Operator	Example	Alternatives
Addition	OR	$a + b$	$a \vee b$
Multiplication	AND	ab	$a \cdot b, a \wedge b$
Complement	NOT	a'	$\overline{a}, \neg a$

Table 9.2 $+$ operation

a	b	$a + b$
0	0	0
0	1	1
1	0	1
1	1	1

Table 9.3 $\cdot$ operation

a	b	$a \cdot b$
0	0	0
0	1	0
1	0	0
1	1	1

- *Associative Laws*:

$$a + (b + c) = (a + b) + c$$
$$a \cdot (b \cdot c) = (a + \cdot b) \cdot c$$

- *Distributive Laws*:

$$a + (b \cdot c) = (a + b) \cdot (a + c)$$
$$a \cdot (b + c) = (a \cdot b) + (a \cdot c)$$

We can replace the boolean operations $+$, $\cdot$ and $'$ by $\vee$, $\wedge$ and $\neg$. The $\wedge$ operation will be evaluated before the $\vee$ operation as in compound logical statements. The notations commonly used are summarized in Table 9.1. Truth tables for $+$, $\cdot$ and $'$ operations are listed in Tables 9.2, 9.3 and 9.4 respectively. We will write ab in short for $a \cdot b$. The following can be derived from the basic laws.

- *Absorbtion*:

$$a + ab = a, \qquad a(a + b) = a$$

since $a + ab = a(1 + b) = a$ and for the dual expression,

Table 9.4 Negation

a	a'
0	1
1	0

$$a(a + b) = aa + ab$$
$$= a + ab = a(1 + b) \qquad \text{Idempotent Law}$$
$$= a$$

- Degenerate Effect

$$a + a'b = a + b \qquad a(a' + b) = ab$$

The first statement can be shown to be valid as follows,

$$a + b = a + b(a + a')$$
$$= a + ab + a'b$$
$$= a(b + 1) + a'b$$
$$= a + a'b$$

and for the dual expression,

$$a(a' + b) = aa' + ab$$
$$= ab \qquad\qquad \text{Complement Law}$$

- *De Morgan's Laws*:

$$(a + b)' = a'b', \qquad (ab)' = a' + b'.$$

9.1.1   Principle of Duality

The laws and theorems of Boolean algebra can be divided into two part as can be observed. *Principle of duality* states that a theorem proved can be proven for one part and the dual of the theorem follows naturally. Dual of such a Boolean expression may be obtained by replacing a "+" operator with "·" operator and vice versa; and replacing 1's with 0's and 0's with 1's. For example, the distributive law states the following,

$$a(b + c) = ab + ac$$

dual of which is $a + bc = (a + b)(a + c)$. Let us prove that this dual expression is valid as below.

Table 9.5 Truth table for Boolean function $a'b + b'c$

a	b	c	a'	b'	$a'b$	$b'c$	$a'b + b'c$
0	0	0	1	1	0	0	0
0	0	1	1	1	0	1	1
0	1	0	1	0	1	0	1
0	1	1	1	0	1	0	1
1	0	0	0	1	0	0	0
1	0	1	0	1	0	1	1
1	1	0	0	0	0	0	0
1	1	1	0	0	0	0	0

$$
\begin{aligned}
(a + b)(a + c) &= aa + ac + ab + bc \\
&= a + ac + ab + bc && \text{Identity Law} \\
&= a(1 + c + b) + bc \\
&= a + bc && \text{Idempotent Law.}
\end{aligned}
$$

9.1.2 Boolean Functions

Definition 9.2 A Boolean expression of the form

$$f(x_1, \ldots, x_n)$$

is called a *Boolean function* of Boolean variables $x_1, \ldots, x_n$.

Example 9.1 Let the Boolean function $f(a, b, c) = a'b + b'c$. The value this function can take for different variable combinations is depicted in Table 9.5.

9.1.3 Sum-of-Products Form

We need standardization of Boolean expressions to evaluate and simplify and possibly implement these expressions. All Boolean expressions can be converted into two forms: sum of products (SOP) or product-of-sums (POS) representations. The SOP representation is simply the sum of terms each of which is a product of logical variables as shown below.

$$f(a, b, c, d) = ab'c + bcd'$$

We have four logical variables a, b, c and d and the SOP form is the sum of two products of these variables. In the standard SOP form, each variable should appear

Table 9.6 Truth table of a logical function

a	b	$f(a, b)$	Product terms
0	0	0	$a'b'$
0	1	1	$a'b$
1	0	0	ab'
1	1	1	ab

in each product term. A nonstandard form can be converted to a standard SOP form as shown in the example below for a logical expression with three logical variables a, b and c.

$$\begin{aligned} f &= ab' + ab'c' + ac' \\ &= ab'(c + c') + ab'c' + a'c(b + b') \\ &= ab'c + ab'c' + ab'c' + a'bc + a'b'c \\ &= ab'c + ab'c' + a'bc + a'bc + a'b'c \end{aligned}$$

We expand a product term of a missing variable by multiplying it by the sum of the normal and inverted missing variable. For example, the first product does not contain c, we then multiply it with $(c + c')$ and since this sum is 1, it will not change the value of the function f.

We have seen how to generate a truth table from a given Boolean expression, we will now consider deriving a logical function from a given truth table. Let us consider the example in Table 9.6 where we have the Boolean function f of two variables a and b.

The last column in this table represents the product terms of a and b where the input literal in this column is negated when the corresponding input value is 0 and not complemented when the input value is 1. The product terms in the last column of this table are named *minterms* where m_i is the minterm for row i. The SOP form of function f is then the sum of minterms where $f = 1$ in this table. Essentially, we are trying to associate input variable values of the function that result in a logical 1 as the function value. For example, when row 1 which is 01 in binary results in $f = 1$, the input combination that can provide this output is $a'b$. We need to sum these minterms because any of these input variable combinations results in a logical value of 1 of the output function. In this particular example,

$$f(a, b) = a'b + ab$$

This expression can be written is summation notation as,

$$f(a, b) = \sum(1, 3)$$

which states that rows 1 and 3 of the function yield a logical 1 value.

Example 9.2 Find the expression for the logical function $f = \sum(2, 3, 5)$ of three variables a, b and c.

Solution: We need to identify the minterms for the corresponding truth table rows first and then work out the minterms as below.

$$2 = 010 = a'bc'$$
$$3 = 011 = a'bc$$
$$5 = 101 = ab'c$$

Thus $f = a'bc' + a'bc + ab'c$. We can see that this expression is not in its simplest form.

$$
\begin{aligned}
f &= a'bc' + a'bc + ab'c \\
&= a'b(c' + c) + ab'c \\
&= a'b + ab'c.
\end{aligned}
$$

$\square$

9.1.4 Product-of-Sums Form

In this form of representation, a boolean function is represented as a product of terms each of which is a sum of the logical variables of the function as in the example blow.

$$f(a, b, c) = (a + b)(a' + b + c')$$

As in SOP form, a standard POS should contain each variable in its terms. In this case, we add the product of the missing variable with itself to the term as shown below. Note that xx' is 0 for a Boolean variable x, thus, the value of the function f is not changed.

$$
\begin{aligned}
f(a, b, c) &= (a + b')(a' + c') \\
&= (a + b' + cc')(a' + bb' + c') \\
&= (a + b' + c)(a + b' + c')(a' + b + c')(a' + b' + c')
\end{aligned}
$$

We will now consider deriving POS form from a truth table. Let us consider the truth table for two variables a and b and an output function f as show in Table 9.7. A row in the last column in this table displays the sum of input literals. A term in a row is obtained by complementing the input value when this value is 0 and not complementing it when its value is 0. This column represents all input combinations. and the terms obtained using this method are called *maxterms*. A maxterm M_i is basically the complement of the minterm m_i for the same input combination, for example, $m_2 = ab'$ and $M_2 = (ab')' = a' + b$. Based on this observation, if we find the product of all maxterms that make the output function $f = 0$, we find the POS representation of the function since $f = 1$ and $f' = 0$ are equivalent based on De Morgan's law.

This time we need the product of maxterms since any maxterm that results in 0 will make the function 0 since we will include it in the product. Note that minterms results in a 1 and maxterms result in a 0 as the output. Thus, representing a function of three variables as $f = \sum(2, 3, 6)$ or $f = \Pi(0, 1, 4, 5, 7)$ are equivalent.

Table 9.7 Truth table of a logical function

a	b	f	Summation terms
0	0	0	$a + b$
0	1	1	$a + b'$
1	0	0	$a' + b$
1	1	1	$a' + b'$

9.1.5 Conversions

A Boolean function in POS form can be converted to a SOP form simply by multiplication to result in the sum of products of the terms. For example,

$$
\begin{aligned}
f(a, b, c) &= (a + b + c')(a' + c)(b + c) \\
&= (aa' + ac + a'b + bc + a'c' + cc')(b + c) \\
&= (ac + a'b + bc + a'c')(b + c) \\
&= abc + a'bb + bbc + a'bc + acc + a'bc + bcc + a'c'c \\
&= abc + a'b + bc + a'bc + ac + bc \\
&= bc(a + 1 + a' + 1) + a'b + ac \\
&= bc + a'b + ac
\end{aligned}
$$

Conversion from SOP to POS form can be done by applying De Morgan's law twice. For example,

$$
\begin{aligned}
f(a, b, c) &= a'b + abc' + ac \\
f'(a, b, c) &= (a'b + abc' + ac)' \\
&= (a'b)'(abc')'(ac)' \\
&= (a + b')(a' + b' + c)(a' + c') \\
&= (aa' + ab' + ac + a'b' + b'b' + b'c)(a' + c') \\
&= (ab' + ac + a'b' + b' + b'c)(a' + c') \\
&= (ab' + ac + a'b' + b')(a' + c') \\
&= aa'b' + aa'c + a'a'b' + a'b' + ab'c' + acc' + a'b'c' + b'c' \\
&= a'b' + ab'c' + a'b'c' + b'c' \text{ (using } x(1 + y) = x) \\
&= a'b' + b'c' \\
f(a, b, c) &= (a'b' + b'c')' \\
&= (a'b')'(b'c')' \\
&= (a + b)(b + c).
\end{aligned}
$$

9.1.6 Minimization

Let f_1 and f_2 be two functions in SOP forms. We can state that f_1 is simpler than f_2 when f_1 has less terms than f_2 and it has no more literals than f_2; or f_1 has less literals than f_2 and it has no more terms than f_2. Whenever there is no function that simpler than a Boolean function f, we say that f is in *minimal form* or just *minimal*.

Definition 9.3 (*prime implicant*) An implicant is a minterm/product in SOP form or a maxterm/sum in POS form. A fundamental product E is called a prime implicant of a Boolean function f if E does not contain any fundamental product with this property.

For example given a function $f = ab + bc$, abc is not a prime implicant of this function as shown below, however, ab and bc are prime implicants.

$$f(a, b, c) = ab + bc$$
$$f(a, b, c) + abc = ab + bc + abc$$
$$= ab(1 + c) + bc$$
$$= ab + bc$$

Karnaugh Maps

Karnaugh maps (K-maps) method is used to find commonly a minimal disjunctive form for a Boolean expression. This technique aims to find all prime implicants of a disjunctive form and the minimal expression is then made equal to the sum of all these prime implicants. A K-map is shown as a matrix with adjacent entries differing by one bit position only. This way, any adjacent entries with a 1 can be reduced to a simpler entry, resulting in a simpler expression. For example, if abc and abc' are both represented in the map, these two entries can be reduced to ab only since $abc + abc' = ab(c + c') = ab$. Prime implicants in a K-map are the largest groups of 1's and any function that is represented as the sum of its prime implicants is in its simplest form.

Two Variable K-maps

A two-variable K-map may be formed as in Fig. 9.1. We need to make sure that each cell in the table differs by one-bit only, thus, minterms are placed accordingly.

A different way of expressing a K-map of two variables a and b is shown in Fig. 9.2a with the boxes filled directly with minterm literals of the function. The general idea of simplification using a K-map is to group as many adjacent 1's as possible in powers of 2. For any such group, any difference bits corresponding to the variable can then be omitted.

Each group in a K-map corresponds to a product term. A Boolean function $f(a, b) = a'b' + ab' + ab$ can be represented in such a K-map by writing minterms that have a value of 1 in the corresponding cells. Note that 0s are not written for simplicity since we will not use them for simplification in this case. The minterms

Fig. 9.1 2-variable K-map

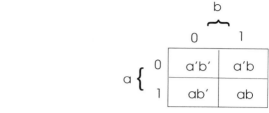

Fig. 9.2 **a** A two-variable K-map with minterms **b** a function

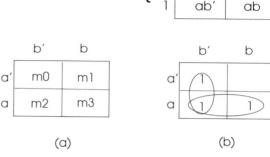

(a)　　　　　　　　　(b)

that have a value of 1 for this function are m_0, m_2 and m_3 which are inserted as 1s in the K-map of Fig. 9.2b. We then attempt to cover these ones in groups of sizes of powers of 2 and try to have these groups as large as possible. Overlapping of groups are possible when they result in more 1s covered. We have two such groups and any literal that exists in normal and converted form in a group can be discarded when forming the product term for the group. For the vertical group, variable a can be discarded as it appears in normal and negated forms in this group resulting in term b', and the horizontal group has the variable b in both forms resulting in representation of this group by variable a. The function f is then $a + b'$ as the sum of these two groups. Had we written this function without using a K-map,

$$f(a, b) = \sum(0, 2, 3)$$
$$\begin{aligned} f(a, b) &= a'b' + ab' + ab & \text{conversion to minterms} \\ &= b'(a + a') + ab & \text{reverse distribution} \\ &= b' + ab & \text{idempotent law} \\ &= a + b' & \text{degenerative effect} \end{aligned}$$

we would arrive at the same simplified function but with some Booelan algebra calculations. In general, the following steps should be considered during simplification of Boolean functions using K-maps.

- We need to generate as few groups as possible to minimize the number of products in the SOP form.
- Each group should be as large as possible to minimize the number of variables in its corresponding product term.
- All minterms should be covered.
- Groups may overlap as long as overlapping increases their sizes.
- Minterms may be covered in different ways resulting in different SOP forms.

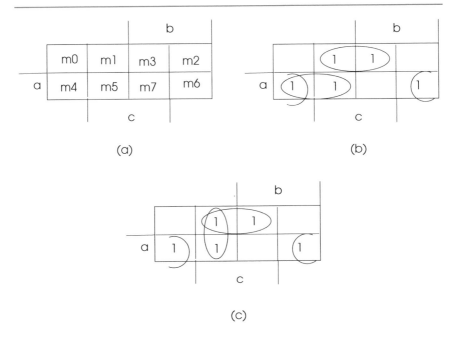

Fig. 9.3 **a** A three-variable K-map with minterms **b** An example function and its simplification **c** Another simplification of the same function

Three Variable K-maps

A three-variable K-map is displayed in Fig. 9.3a for three Boolean variables a, b and c with its minterms. Let us consider simplifying a function $f = \sum(1, 3, 4, 5, 6)$ using a three-variable K-map. We first fill all minterms in the table as shown in Fig. 9.3b and then group 1's in as large as possible groups of sizes of power of 2 which is displayed.

Note that wrapping around the corners of the table is possible since these minterms differ in one bit position only. The upper horizontal group is $a'c$ since these two variables do not change in this group, the lower horizontal group is ab' and the wrapped corner group is ac' using similar reasoning. The function f is the sum of these groups as below,

$$f = a'c + ab' + ac'$$

Let us check using laws of Boolean algebra as follows,

$$f = \sum(1, 3, 4, 5, 6)$$
$$f = a'b'c + a'bc + ab'c' + ab'c + abc'$$
$$f = a'c(b + b') + ab'(c + c') + abc' \qquad \text{reverse distribution}$$
$$f = a'c + ab' + abc' \qquad\qquad\qquad \text{idempotent law}$$
$$f = a'c + a(b' + bc') \qquad\qquad\qquad \text{reverse distribution}$$
$$f = a'c + a(b' + c') \qquad\qquad\qquad\; \text{degenerative effect}$$
$$f = a'c + ab' + ac'$$

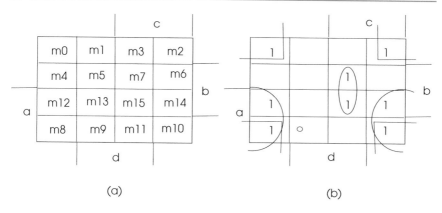

Fig. 9.4 **a** A four-variable K-map, **c** Representation of a function

to result in the same simplified function. If we cover the groups as in Fig. 9.3c, we have the following expression,

$$f = a'c + b'c + ac'$$

which shows us different coverings may result in different but equivalent simplified forms of a function. The equivalence of two expressions may be checked by a truth table.

Four Variable K-Maps

A four variable K-map with minterms is displayed in Fig. 9.4a. As with 2 and 3 variable K-maps, we need to ensure any minterm placed differs by 1 bit with any adjacent minterm. Let us consider implementing $f = \sum(0, 2, 7, 15, 8, 10, 12, 14)$ using a four-variable K-map. Grouping of 1's is shown in Fig. 9.4b and we can generate function f in its simplest form based on these groups as follows. Note that a group with 4 minterms eliminates two variables and a group with 2 minterms eliminates one variable.

$$f = ad' \text{ (wrapped side)} + bcd \text{ vertical } + b'd' \text{ (wrapped corners)} .$$

9.2 Combinational Circuits

A combinational circuit has a number of inputs and its output is uniquely defined for each input combination. Such a circuit has no memory; in other words, its output is dependent only on its current input values but not on any previous input or output values. A combination circuit is constructed using basic building blocks called *gates*.

Fig. 9.5 **a** 2-input *AND* gate, **b** 2-input *OR* gate, **c** *NOT* gate

(a)   (b)   (c)

Table 9.8 Truth table for *OR* and *AND* gates

a	b	a + b	a · b
0	0	0	0
0	1	0	1
1	0	0	1
1	1	1	1

Table 9.9 Truth table for *NOT* gate

a	a'
0	1
0	0

Table 9.10 NAND-OR equivalence

a	b	a'	b'	(ab)'	a' + b'
0	0	1	1	1	1
0	1	1	0	1	1
1	0	0	1	1	1
1	1	0	0	0	0

9.2.1 Gates

The first gate we will consider is the *AND* gate. It has two or more bit inputs and its output is formed by logically *and*ing of these inputs as depicted in Fig. 9.5a. The second basic gate that can be used as the building block of more complicated combinational circuits is the *OR* gate shown in Fig. 9.5b. The output of an *OR* gate is logical 1 whenever one or both of the input bits is a 1. The *NOT* gate has one input bit and simply negates its input as shown in Fig. 9.5c. The truth tables for these basic gates against possible input bit combinations are depicted in Tables 9.8 and 9.9.

Variations of these basic gates are the *NAND* gate, inverted input *AND* gate and inverted input *OR* gate. The *NAND* gate is basically an *AND* gate with an inverted output as shown in Fig. 9.6a. Applying De Morgan's law for this gate with binary inputs a and b yields the following,

$$(ab)' = a' + b'$$

which means this gate is basically equivalent to an inverted input *OR* gate. This fact is depicted in Table 9.10 by the equality of the last columns.

Fig. 9.6 **a** 2-input NAND gate, **b** 2-input NOR gate, **c** 2-input XOR gate

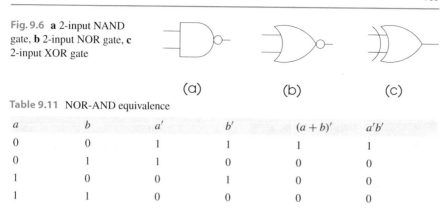

(a) (b) (c)

Table 9.11 NOR-AND equivalence

a	b	a'	b'	$(a+b)'$	$a'b'$
0	0	1	1	1	1
0	1	1	0	0	0
1	0	0	1	0	0
1	1	0	0	0	0

Table 9.12 Truth table for *XOR* and *XNOR*

a	b	$a \oplus b$	$(a \oplus b)'$
0	0	0	1
0	1	1	0
1	0	1	0
1	1	0	1

Fig. 9.7 **a** 2-input *XNOR* gate gate **b** 2-input inverted input AND gate, **c** 2-input inverted input OR gate

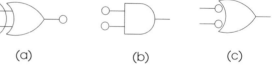

(a) (b) (c)

A *NOR* gate is basically an inverted output *OR* gate as shown in Fig. 9.6b. Applying De Morgan's law for this gate results in the following,

$$(a + b)' = a' \cdot b'$$

which shows that a *NOR* gate is equivalent to an inverted input *AND* gate as verified by the truth table of Table 9.11.

The *XOR* (exclusive-OR) gate has a similar functionality to what we saw for the compound propositions in Chap. 1, its output is 1 only when the inputs are not the same. The symbol for this gate is shown in Fig. 9.6c, we will show the output of a 2-input *XOR* gate with inputs a and b as $a \oplus b$. The *XNOR* gate is basically an *XOR* gate with an inverted output as shown in Fig. 9.7a where an inverted input AND and inverted input OR gates are shown in (b) and (c) of the same figure. The truth tables for *XOR* and *XNOR* gates with two inputs are shown in Table 9.12.

Fig. 9.8 **a** $ab + bc$,
b $ab(c + d)$

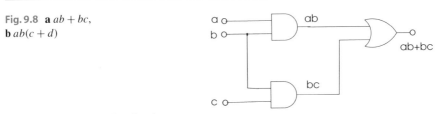

Table 9.13 Truth table for $ab + bc$

a	b	c	ab	bc	$ab + bc$
0	0	0	0	1	1
0	0	1	1	0	1
0	1	0	0	0	0
0	1	1	0	0	0
1	0	0	0	1	1
1	0	1	0	1	1
1	1	0	1	0	1
1	1	1	1	1	1

9.2.2 Designing Combinational Circuits

Using these basic gates, we can build more complicated circuitry. We will first consider two-stage circuits in which the output from the first stage is formed in parallel
and fed to the second stage which then produces the output. The output is stabilized
only after the output from the first stage is stabilized. We have two basic forms of
two stage circuits; *AND-OR* circuits and *OR-AND* circuits.

9.2.2.1 AND-OR Circuits

Consider the circuit of Fig. 9.8 where 3 inputs a, b and c are fed to this circuit. The
first stage of the circuit has two *AND* gates which work in parallel to form the outputs
ab and bc. The second stage of this circuit has an *OR* gate which adds these two
inputs to get $ab + bc$. Note the use of a dot to indicate that there is a connection
when two wires cross, otherwise, dots are omitted. The truth table for this may be
formed as in Table 9.13. This type of arrangement is denoted by *AND-OR* circuitry
and represents *SOP* form of a boolean expression conveniently.

Clearly, when constructing such a circuit, we need to have the SOP form in its
simplest form. We can have an algorithm to build a logic circuit for a given boolean
function consisting of the following steps.

Table 9.14 Truth table for f_1 and f_2

a	b	c	f_1	f_2
0	0	0	0	1
0	0	1	0	0
0	1	0	0	0
0	1	1	1	0
1	0	0	0	1
1	0	1	0	1
1	1	0	1	0
1	1	1	1	1

1. Express the boolean function in SOP form which can be achieved by forming the truth table of the expression and finding the sum of minterms where the output is a logical 1.
2. Simplify the expression using K-maps or any other method such as using laws of and theorems of Boolean algebra.
3. Construct the logic circuit in *AND-OR* stages.

Example 9.3 Consider the truth table of two Booelan functions f_1 and f_2 of three variables a, b and c shown in Table 9.14. Construct the logical circuit to implement functions f_1 and f_2.

Solution: Note that we have two output functions to consider. Let us first form the SOP expressions for f_1 and f_2 using the minterms.

$$f_1 = \sum(3, 6, 7) = a'bc' + ab'c' + ab'c + abc$$

Simplification using laws of Boolean algebra yields,

$$f_1 = a'bc + ab(c + c')$$
$$= a'bc + ab$$
$$= b(a + a'c')$$
$$= b(a + c')$$
$$= ab + bc'$$

Similarly,

$$f_2 = \sum(0, 4, 5, 7) = a'b'c' + ab'c' + ab'c + abc$$
$$= b'c'(a + a') + ac(b + b')$$
$$= b'c' + ac$$

The *AND-OR* circuit to realize functions f_1 and f_2 is depicted in Fig. 9.9.

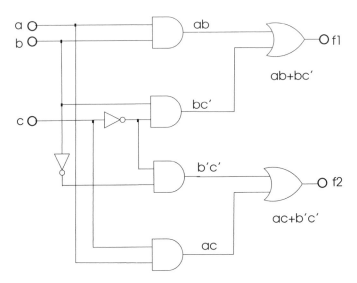

Fig. 9.9 Realization of f_1 and f_2 of Example 9.3

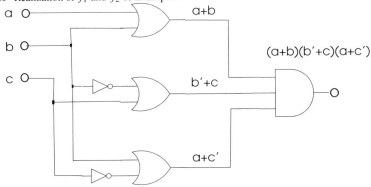

Fig. 9.10 OR-AND circuit example

9.2.2.2 OR-AND Circuits

This type of circuitry is characterized by a number of *OR* gates in the first stage and the outputs of these gates are fed to a multiple input *AND* gate as depicted in Fig. 9.10. There are 3 logical inputs a, b and c and the output is the product of the combinations of these sums. The truth table of the function represented by this combinational circuit is shown in Table 9.15.

Example 9.4 Find the expression for the logical function $f = (a+b+c')(a'+b+c)$.
Solution: We can implement this function directly from this expression and the resulting circuit in POS form is depicted in Fig. 9.11.

Table 9.15 Truth table for $ab + bc$

a	b	c	$a+b$	$b'+c$	$a+c'$	f
0	0	0	0	1	1	0
0	0	1	1	1	0	0
0	1	0	1	0	1	0
0	1	1	1	1	0	0
1	0	0	1	1	1	1
1	0	1	1	1	1	1
1	1	0	1	0	1	0
1	1	1	1	1	1	1

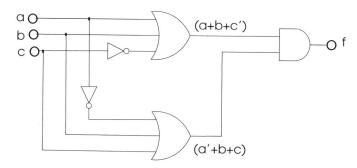

Fig. 9.11 Realization of f of Example 9.4

However, simplifying the expression results in a simpler circuit in general. Thus,

$$f = (a + b + c')(a' + b' + c)$$
$$= aa' + ab + ac + a'b + bb + bc + a'c' + bc' + cc'$$
$$= b(a + a') + b(c + c') + b + ac + a'c'$$
$$= b + ac + a'c'$$

This simplification provided SOP form. Double negation provides POS form in the form of a complemented output as below.

$$f = b + ac + a'c'$$
$$f' = (b + ac + a'c')'$$
$$= b' \cdot (ac)' \cdot (a'c')'$$
$$= b' \cdot (a' + c') \cdot (a + c)$$
$$f = (b' \cdot (a' + c') \cdot (a + c))'$$

The circuit to implement this simplified POS form of the expression as *OR-AND* gate configuration is depicted in Fig. 9.12. We have 2-input *OR* gates in this realization

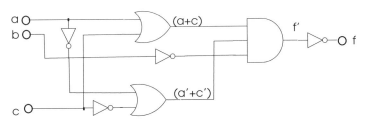

Fig. 9.12 Realization of f of Example 9.4

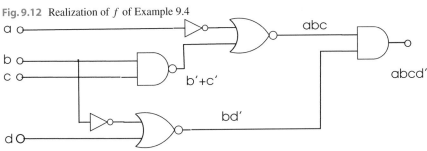

Fig. 9.13 A multistage combinational circuit

of the circuit which are simpler to construct than 3-input ones, however, we have a 3-input *AND* gate in the second stage now.

9.2.2.3 General Circuit Layout

We can have a logic circuit consisting of several stages with a combination of gates in general. Figure 9.13 displays such a circuit with four inputs a, b, c and d. The final output is the product of the four inputs which means a four-input AND gate will realize the same function. A four-input gate is more difficult to construct than a two or three input one, however, this circuit has three stages excluding the inverters, with each stage contributing to the delay in obtaining a stabilized output.

9.2.3 Arithmetic Circuits

Circuits to perform arithmetic operations on binary data can be formed using simple logic gates. Let us consider adding two n-bit binary numbers by first examining addition of two bits. The truth table for this operation is shown in Table 9.16 where S is the sum and C is the carry from the addition.

A close look at this table reveals that S can be realized by an XOR gate and C by an AND gate as depicted in Fig. 9.14. This circuit is called a *half-adder* since it performs addition of two least significant bits of two n-bit binary numbers. For any

Table 9.16 Half-adder truth table

a	b	S	C
0	0	0	0
0	1	1	0
1	0	1	0
1	1	0	1

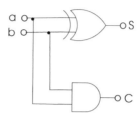

Fig. 9.14 Half-adder circuit

Table 9.17 Truth table for full-adder

a	b	C_{in}	S	C_{out}
0	0	0	0	0
0	0	1	1	0
0	1	0	1	0
0	1	1	0	1
1	0	0	1	0
1	0	1	0	1
1	1	0	0	1
1	1	1	1	1

other bit addition of two binary numbers, we need to input carry bits from lower bit positions. Thus, the *full-adder* performs addition of two bits by taking an input carry bit. The truth table for this operation is depicted in Table 9.17 with C_{in} as input carry bit and C_{out} as the output carry bit.

A full-adder circuit can be realized by cascading two half-adders as shown in Fig. 9.15.

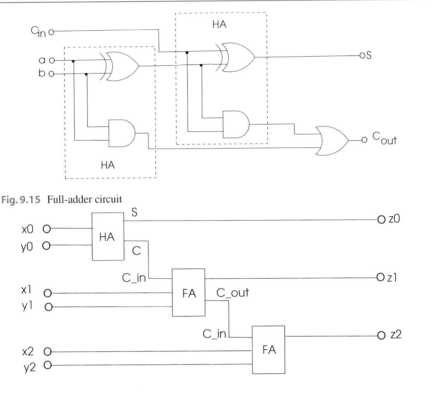

Fig. 9.15 Full-adder circuit

Fig. 9.16 Addition of two n-bit numbers

A logic circuit to add to n-bit numbers x and y is shown in Fig. 9.16. The carry-out bit from each full-adder (FA) stage is fed to the carry-in bit of the next full-adder stage. Note that a half-adder (HA) is used for the least significant bit addition of two numbers since we do not have any preceding stage, thus, there is no carry-in bit input.

9.3 Review Questions

1. What is meat by Boolean algebra?
2. State the idempotent and identity laws of Boolean variables.
3. How is the dual of a Boolean law obtained?
4. Describe the SOP and POS representations of a Boolean function.
5. What is the basic principle of K-maps?
6. What are the basic logic gates?
7. What are the steps of realizing a Boolean function represented in SOP form using logic circuitry?
8. What are the steps of realizing a Boolean function represented in POS form using logic circuitry?
9. What is a half-adder and a full-adder? Where can we use these circuits?

9.4 Chapter Notes

This chapter comprised two related sections: Boolean algebra and combinational circuits. Boolean algebra is defined on a set of binary variables with operations $+$, $\cdot$ and inversion. We reviewed basic laws of Boolean algebra and Boolean functions which may be represented as sum-of-products (SOP) and product-of-sums (POS) forms. K-maps can be used effectively to simplify a Boolean function of 2, 3 and 4 variables.

In the second part of the chapter, we reviewed combinational circuits which provide an output dependent only on the current values of their inputs. We described basic building block of combinational circuits which are OR, AND and NOT gates. Various other gates can be formed using these gates and logic circuits are structured using the basic gates. Logic circuits may be built using AND-OR circuits which are formed by AND gates in the first stage and an OR gate possibly with a number of inputs greater than 2 in the second stage. An AND-OR gate circuitry conveniently represents SOP description of a logical function. A simple procedure to build a logic circuit to represent a Boolean function is to form the SOP expression of the function, to simplify this expression and use AND-OR circuitry that provides the function at its output. A Boolean function may be represented in simplified POS form, and an OR-AND circuitry may be used to implement this function in POS form. Lastly, we reviewed the half-adder circuit which inputs two binary bits and produces their sum and a carry bit output. The full-adder circuit has three bit inputs; two bits are the binary digit values of the number to be added and the third bit is the carry-in bit from a preceding digit of the binary number to be added. Cascaded full adders may be used to add two n-bit binary numbers. In practice, more complicated logic circuits are used to add binary numbers since an n-stage full adder circuit built from half-adders needs delays dependent on n to have the output stabilized. The output of a *sequential* circuit depends on the current value of its inputs and its current state which are the basic building blocks of storage elements in a computer.

Exercises

1. Simplify the following Boolean functions using laws of Boolean algebra.

   a. $A = abc + abc' + a'bc' + a'b'c'$.
   b. $B = ab + ab' + a'bc' + a'b'c'$.

   c. $C = (a + b')(bc') + (ab + c)ac'$.
   d. $D = ab + (ab'c)b + ab'$.

2. Convert the following Boolean function to standard SOP form.

$$f(a, b, c) = a'bc + ac + bc + c'$$

3. Convert the following Boolean function to standard POS form.

$$f(a, b, c) = (a' + b + c)(a + b')(a + b + c')(b + c')$$

4. Convert the following SOP representation of a Boolean function to POS representation.

$$f(a, b, c) = abc' + ab'c + abc + b'c$$

5. Convert the following POS representation of a Boolean function to SOP representation.

$$f(a, b, c) = (a' + b + c)(a + b' + c)(a + b' + c')(a + b + c')$$

6. Work out the SOP form of the function $f(a, b, c) = \sum(1, 3, 5, 6)$ using its minterms and simplify using K-maps.

7. Work out the POS form of the function $f(a, b, c) = \Pi(0, 2, 5, 7)$ using its maxterms and simplify using K-maps.

8. Provide the AND-OR circuit for a function $f(a, b, c) = \sum(1, 2, 5)$ by writing its SOP form first and then simplifying.

9. Provide the OR-AND circuit for a function $f(a, b, c) = \Pi(0, 2, 3, 5)$ by writing its POS form first and then simplifying.

10. Find the sum of two 8-bit binary numbers 1101 1001 and 0101 1110.

11. Design a full adder circuit by forming the truth table for the sum and carry-out bits, drawing the K-maps for these outputs, minimization and drawing the equivalent combinational circuit.

Reference

1. Biography in Encyclopaedia Britannica. http://www.britannica.com/biography/George-Boole

Introduction to the Theory of Computation

10

Theory of computation deals with developing mathematical models of computation. This area of research is divided into three subareas: complexity theory, computability theory and automata theory. We mostly review basic structures of automata theory which are languages and finite state automata in this chapter. A language is defined over a set of symbols called an alphabet. A finite state machine is a mathematical tool to model a computing system. Unlike a combinational circuit, a finite state machine has a memory and its behavior and output depends on its current input and its current state. A finite state automata is a finite state machine with no outputs; instead, it has final states called accepting states. A finite state automata can be used to recognize a language conveniently. We review languages, finite state machines, finite state automata, language recognition in this chapter and conclude the first part with the Turing machine named after Alan Turing, which is a more general type of a finite state machine. We then have a short review of complexity theory with the basic complexity classes.

10.1 Languages

An *alphabet* Σ is a finite set of symbols such as $\{a, b, \ldots, z\}$ or $\{0, 1\}$. A string over an alphabet Σ is a finite sequence of symbols of Σ. For example, 1101001 is a string over the alphabet $\Sigma = \{0, 1\}$ and $accbba$ and $babaaac$ are strings over the alphabet $\Sigma = \{a, b, c\}$. The set of all strings over an alphabet Σ is denoted by Σ^* and the empty string is denoted by ε.

A string w over an alphabet Σ is shown by $w \in \Sigma^*$ since w is generated from the symbols of Σ and thus should be a subset of the set Σ^*; the length of w is shown by $|w|$. The set of all strings with length n over an alphabet Σ is denoted by Σ^n. For example, let $\Sigma = \{a, b\}$, then $\Sigma^2 = \{aa, ab, bb, ba\}$.

© Springer Nature Switzerland AG 2021
K. Erciyes, *Discrete Mathematics and Graph Theory*, Undergraduate Topics in Computer Science, https://doi.org/10.1007/978-3-030-61115-6_10

Formally, a *language* L is a subset of the strings in Σ^* shown by $L \in \Sigma^*$. A language can be specified using the set builder notation; for example, if $\Sigma = \{0, 1\}$, then the following define various languages.

- $L_1 = \{w||w| = 2\}$. Based on this definition, $L_1 = \{00, 01, 10, 11\}$
- $L_2 = \{w||w| = 3 \text{ and } w \text{ ends with "11" }\}$. Thus, $L_2 = \{011, 111\}$
- $L_3 = \{w||w| = 4 \text{ and } w \text{ contains "10" as a substring}\}$. Therefore, $L_3 = \{0010, 0100, 0101, 0110, 1010, 1100, 1101, 1110\}$

Let us define the following operations on two languages $L_1, L_2 \in \Sigma^*$ by letting $L_1 = \{a, bc, c\}$ and $L_2 = \{a, b, cc\}$ for all examples.

- *Union*: The union of languages L_1 and L_2 is the set union operation. For example, $L_1 \cup L_2 = \{a, b, c, bc, cc\}$.
- *Concatenation*: This operation is like the cartesian product of two sets without the commas in elements. Formally,

$$L_1 \circ L_2 = \{xy \in \Sigma^* | x \in L_1, y \in L_2\}$$

For example, $L_1 L_2 = \{aa, ab, acc, bca, bcb, bcc, ca, cb, ccc\}$. The concatenation symbol $\circ$ is frequently omitted so that $L_1 \circ L_2$ is written as $L_1 L_2$.
- *Star*: The star of a language L is denoted by L^* which contains *all* words that can be broken down into words that are included in L. For example, $L_1^* = \{a, aa, a \ldots a, abc, aabcc, \ldots\}$

We are now ready to define a regular language

Definition 10.1 (*regular language*) The set of regular languages over an alphabet let Σ is defined as follows.

- The empty set $\emptyset$ is a regular language.
- The language consisting of the empty string ε is a regular language.
- Any symbol $a \in \Sigma$ is a regular language.
- If L_1 and L_2 are regular languages, $L_1 \cup L_2, L_1 L_2, L^*$ and L_2^* are regular languages.

Based on this recursive definition of a regular language, we can generate a regular language starting from the alphabet and using the operations described.

10.2 Finite State Machines

A combinational circuit we have reviewed in Chap. 9 has no memory, the output of such a circuit depends on its current inputs only. A *finite state machine* (FSM) is a mathematical model for representing a computing system with little memory. Its behavior depends on its current inputs and its current state. For example, ringing of a

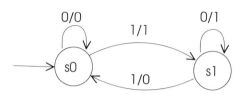

Fig. 10.1 FSM diagram of an odd parity checker

bell at the end of a class in a school means taking a break but the same ringing at the end of the break means going in the classroom. Thus, the same input may produce a different output depending on the current state of the system in a FSM.

Definition 10.2 (*finite state machine*) An FSM M is a 5-tuple:

- I: A finite set of inputs (input alphabet)
- S: A finite set of states
- $O \subset S$: A finite set of outputs (output alphabet)
- $s_0 \in S$: An initial state in S
- $F : S \times I \rightarrow S$: A transition function that maps the cartesian product of S and I to S.

An FSM is commonly represented by a diagram with circles representing the possible states and arcs showing the transitions between the states. This representation is a directed graph as we will review in Part II. Each arc is labeled with possible inputs and the produced outputs separated by "/". Let us consider an FSM that inputs binary strings and counts the numbers of 1's in an input string. If the current count is an odd number, the FSM outputs 1 otherwise its output is 0. This machine is called an *odd parity checker* and may be used for error detection in digital transmissions. The state diagram of this FSM is depicted in Fig. 10.1.

The definition parameters for this FSM are as follows: $I = \{0, 1\}$, $S = \{s_0, s_1\}$, $O = \{0, 1\}$, the initial state is s_0 and the transition function F is $(s_0, 0) \rightarrow s_0$, $(s_0, 1) \rightarrow s_1$, $(s_1, 0) \rightarrow s_1$, $(s_1, 1) \rightarrow s_0$. Let us consider an input sequence of 0110110 from left to right to this FSM. The state changes will be $s_0, s_1, s_0, s_0, s_1, s_0, s_0$ respectively.

Example 10.1 Let us consider a vending machine that provides chocolates worth 20 cents. For simplicity, we will assume it accepts 5 or 10 cent coins only. It has two outputs: Release (R) to release chocolate and Change (C) to give back change. The FSM diagram of this machine with four states is depicted in Fig. 10.2. States A, B, C and D correspond to 0, 5, 10, and 15 cent states and the output in negated form such as R' or C' means that output is not activated. Note that we do not need a 20 cent state as depositing such amount means we need to get back to state A after releasing the chocolate.

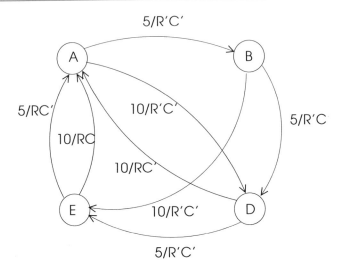

Fig. 10.2 FSM diagram for vending machine

Table 10.1 FSM table for the vending machine

	5	10
A	*B, R′C′*	*D, R′C′*
B	*D, R′C′*	*E, R′C′*
D	*E, R′C′*	*A, RC′*
E	*A, RC′*	*A, RC*

Let us now form the FSM table with 4 states and 3 inputs based on this diagram. The entries in the table show the next state and the output. The table representation of an FSM is convenient for coding the FSM. Let us denote the states of the vending machine by 1, 2, 3 and 4 and the 5 and 10 cent inputs by 1 and 2. The FSM table may be represented by an array and each array entry of Table 10.1 can be assigned to the address of a procedure that will perform the required action. For example, procedure $p[1, 1]$ needs to assign current state to B (2) and set outputs $R = 0$, $C = 0$. The algorithm for the FSM then consists of few lines of code as shown in Algorithm 10.1. The FSM array (FSM_tab) has n_states rows which is the number of states, and n_inputs which is the number of inputs. In the vending machine example, these parameters are 4 and 2 respectively.

Algorithm 10.1 FSM algorithm

1: $FSM\_tab[n\_states, n\_inputs] \leftarrow$ addresses of procedures
2: $current\_state \leftarrow 0$
3: $sum \leftarrow 0$
4: **input** num
5: **while** $true$ **do**
6:    **input** in
7:    call the procedure in $FSM[current\_state, in]$
8: **end while**

10.3 Finite State Automata

A finite state automaton (FSA) is an FSM without any output but with a set of terminating states called the *accepting states*.

Definition 10.3 (*finite state automaton*) A finite state automaton (FSA) F is a 5-tuple:

- Σ: A finite alphabet,
- S: A finite set of states,
- $s_0 \in S$: An initial state in S
- $\delta : S \times \Sigma \rightarrow S$: A transition function,
- $F \subset S$: A finite set of accepting states

The automaton M is expressed as $M(I, S, Y, s_0, F)$. Accepting states which are denoted by double circles are the places that FSM stops running.

10.3.1 Analysis

Let us consider the FSA M of Fig. 10.3. The parameters for M are as follows: $\Sigma = \{0, 1\}$, $S = \{s_0, s_1, s_2\}$, $A = \{s_2\}$, the initial state is s_0 and the transition function δ is $(s_0, 0) \rightarrow s_0$, $(s_0, 1) \rightarrow s_1$, $(s_1, 0) \rightarrow s_2$, $(s_1, 1) \rightarrow s_1$, $(s_2, 0) \rightarrow s_1$, $(s_2, 1) \rightarrow s_2$. Note that we do not have any outputs. When we trace possible inputs, it can be seen that M reaches the accepting state for inputs 010, 10, 110, 11110 or any input that contains 10 consecutively. Thus, it detects 10 sequence by reaching its accepting state. When the sequence 01111 is input to this FSA, its final state is s_1 which is not its accepting state. When this happens, we say M *rejects* 01111.

The FSA table for this automaton is displayed in Table 10.2 with each entry showing the next state of the FSA when the input shown in the columns is received.

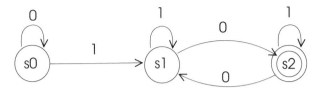

Fig. 10.3 A finite state automata

Table 10.2 State table for FSA of Fig. 10.3

	0	1
s_0	s_1	s_1
s_1	s_1	s_1
s_2	s_1	s_1

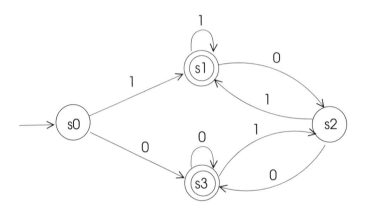

Fig. 10.4 A finite state automaton

Example 10.2 Consider the state diagram of the FSA M of Fig. 10.4. The parameters are $\Sigma = \{0, 1\}$, $S = \{s_0, s_1, s_2, s_3\}$, $F = \{s_1, s_3\}$ and the initial state is s_0. It has two accepting states as s_1 and s_3 shown by double circles.

The state table for M is depicted in Table 10.3. A close look at how M behaves reveals that inputs 0, 00, 010, …which start with a 0 and end with a 0 cause it to reach the accepting state s_3. Similarly, input strings 1, 11, 101, …that start with a 1 and end with a 1 are accepted at state s_1. Note that this FSA will not accept input strings 1000, 0101 which begin and end with different symbols.

Table 10.3 State table for FSA of Fig. 10.4

	0	1
s_0	s_3	s_1
s_1	s_2	s_1
s_2	s_3	s_1
s_3	s_3	s_2

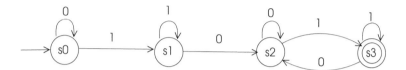

Fig. 10.5 A finite state automaton

Table 10.4 State table for FSA of Fig. 10.5

	0	1
s_0	s_0	s_1
s_1	s_2	s_1
s_2	s_2	s_3
s_3	s_2	s_3

10.3.2 Designing Finite State Automata

Let us now consider working the other way around, given a pattern that an FSA M accepts, how can we design M? Designing M means forming the state diagram and possibly the state table in this sense. Consider an FSA M that should accept a binary string which starts with a binary 1 and ends with a binary 1 and has at least one 0 in between the first and last symbols. We can draw the state diagram of M as in Fig. 10.5.

Tracing through the states of M, we can see that it waits for at least one 1 to move to state s_1, and it gets locked at that state until a 0 symbol is received in which case it moves to state s_2. This time it waits for a 1 to reach the accepting state of s_3. Some example input strings accepted by M are 101, 1101, 11000111, ... The state table for M is depicted in Table 10.4.

Fig. 10.6 An example FSA

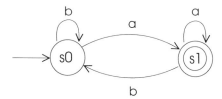

10.4 The Relationship Between Languages and Automata

An FSA *accepts* an input string if the input string tracing through the FSA leaves the FSA in an accepting state. The set of strings accepted by an FSA M is denoted by the language $L(M)$ accepted by that FSA or the language of the FSA. Note that $L(M) \subseteq \Sigma^*$. A formal definition of the language of a FSA is as follows.

Definition 10.4 (*language*) Let $M = (\Sigma, S, s_0, \delta, F)$ be a FSA and let $w = w_1 w_2 \ldots w_n$ be a string over I. A sequence of states $q_0, q_1, \ldots, q_n$ are defined as follows.

- $q_0 = s_0$,
- $q_{i+1} = F(q_i, w_{i+1})$, for $i = 0, 1, \ldots, n - 1$.

If $q_n \in A$, then M accepts w, if $q_n \notin A$, then M rejects w.

Informally, if M arrives at any of the accepting states at the end of the input string, M accepts w. If none of the states are reached at the end of the input w, M rejects w. Now, we have an alternative definition of a regular language.

Definition 10.5 (*regular language*) A language is called a *regular language* if there exists a finite automaton that recognizes it.

Let us consider the simple FSA M of Fig. 10.6 with $\Sigma = \{a, b\}$, $S = \{s_0, s_1\}$, $F = \{s_1\}$ and s_0 as the initial state.

The state table for this FSA is shown in Table 10.5. Some of the possible inputs to reach the accepting state s_1 are ba, bba, $bbba$; in fact any input string that ends with a a. Thus, we can specify the language for M as follows,

$$L(M) = \{w | w \text{ ends with } a\}$$

By the definition of a regular language, in order to prove that a language is regular, we need to find an FSA M that accepts it. Let us attempt to design a FSA M that accepts a binary string that contains 010 as a substring. We note the following while considering the design:

- M should stay in the initial state until the first 0 is encountered.

Table 10.5 State table for FSA of Fig. 10.3

	a	b
s_0	s_1	s_0
s_1	s_1	s_0

Table 10.6 State table for FSA of Fig. 10.3

	a	b
s_0	s_1	s_0
s_{0*}	s_1	s_{01}
s_{01}	s_{010}	s_0
s_{010}	s_{0*}	s_0

Fig. 10.7 The FSA that accepts 010 as a substring

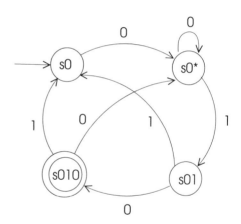

- When the first 0 is detected, it should go to state s_{0*} meaning a 0 is detected.
- When a 1 is detected in state s_{0*}, it should go to state s_{01} to mean a 01 in sequence is detected.
- When a 0 is detected in state s_{01}, it should arrive at the receiving state s_{010} to mean a 01 in sequence is detected.

Note that we walked through the legitimate states to reach the accepting state. The state table based on the foregoing logic is depicted in Table 10.6. We can now form the state diagram of M as shown in Fig. 10.7.

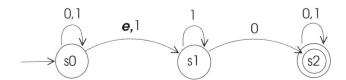

Fig. 10.8 Nondeterministic finite state automata example

Table 10.7 State table for FSA of Fig. 10.8

	0	1	ε
s_0	s_0	$\{s_0, s_1\}$	s_1
s_1	s_2	s_1	–
s_2	s_2	s_2	–

10.5   Nondeterministic Finite State Automata

The finite state automata we have reviewed until now works by receiving an input and moving to the next state based on the value of the input. The next state accessed is distinct and thus, this type of automata is termed *deterministic* finite state automata (DFSA). Nondeterministic finite state automata (NFSA) is basically a finite state automata with one important difference; the next state when an input is received may be more than one state. Consider the NFSA of Fig. 10.8, notable differences from a DFSA are as follows: transition from state s_0 when a 1 is received is either to state s_1 or to s_0; the existence of an input ε label of transaction from set s_0 to s_1 means this transition may take without any input and the accepting state has no outward transitions meaning once NFSA reaches that state, it stays there. Let us form the state table for this NFSA as in Table 10.7. Note the use of ε as an input and next states consisting of more than one state also, when an input is not allowed at a state, we have "–" symbol or not applicable (NA) entry in a state table. An important question to be answered in such an automaton is how it will proceed, for example, when it receives a 1 at state s_0. The basic operation of a NFSA M is performed by the following steps [1].

- When there are more than one choice, M copies itself into multiple copies and executes all of these copies in parallel.
- Each copy of M when confronted with multiple choices, performs as in the previous step.
- When the next input at a state s in a parallel running copy of M is not contained in any of the transactions from s, that copy is terminated along with the branch that contains it.
- If any of the running copies of M reaches an accepting state at the *end* of the input string, NFSA accepts the string. Notethat there should be at least one such branch to have M accept the input string.

Fig. 10.9 Tree diagram for
the NFSA of Fig. 10.8

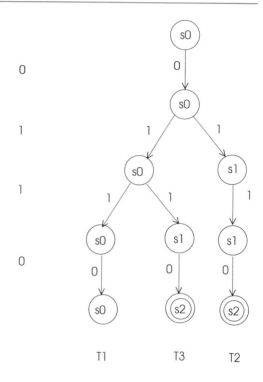

Let us implement this procedure to the NFSA M of Fig. 10.9 using a tree showing
all branches of possible parallel executions starting from the initial state when a
binary string 0110 is received.

The automaton M starts running as the main thread T_1 from the initial s_0 state
and when the first symbol 0 is received, it stays at this state. The second symbol is a
1 which causes M to replicate itself and start another thread T_2. Receiving another
1 for T_1 causes it to replicate itself to thread T_3 and T_2 stays in state s_1 with this
input. When the final symbol 0 is received, T_1 stays at state s_0, T_2 and T_3 both arrive
at the accepting state s_2 and M terminates. Thus, the input 0110 is accepted by M,
however, it can be seen that input string 0111 will not be accepted by M as we need a
final 0 to reach state s_2. A closer inspection reveals that M accepts any binary string
that starts and ends with a 0 and has at least one 1 in between the 0s. We can now
define a nondeterministic finite state automaton formally as follows.

Definition 10.6 (*nondeterministic finite state automaton*) A nondeterministic finite
state automaton (FSA) F is a 5-tuple:

- Σ: A finite alphabet,
- S: A finite set of states,
- $s_0 \in S$: An initial state in S
- $\delta : S \times \Sigma_\varepsilon \rightarrow \mathcal{P}$: A transition function,
- $F \subset S$: A finite set of accepting states

where $\Sigma_\varepsilon = \Sigma \cup \{\varepsilon\}$. Note that the differences from a DFSA are the transition function may take the empty string ε as the input and it may transfer the state of the machine to a number of states, not only to one distinct state as in DFSA. DFSAs and NFSAs are equivalent as they both support the same class of languages [1]. The class of regular languages is closed under the union and star operations.

10.6 Regular Expressions

A *regular expression* is a statement constructed from regular operations. Consider the regular expression,

$$(0 \cup 1) \circ 1^*$$

The value of this expression is a language that contains all strings that start with 0 or 1 followed by any number of 1s, for example 01111 or 1111111. We have the basic concatenation ($\circ$) which is commonly omitted and star (*) regular operations used in this expression. A formal definition of a regular expression is as follows.

Definition 10.7 (*regular expression*) A regular expression R can be any of the following,

- $a \in \Sigma$
- ε,
- $\emptyset$
- The union of two regular expressions,
- The concatenation of two regular expressions,
- The concatenation of a regular expression.

Some examples of regular expressions over the alphabet $\Sigma = \{0, 1\}$ and the language L they describe are as follows.

- 0^*: L does not contain any 1s.
- $(0 \cup 1)(0 \cup 1)$: L contains all strings of length 2. $L = \{00, 01, 10, 11\}$. Note that L is finite in this case.
- 0^*10^*: L contains a single 1, that is, $L = \{010, 0010, 01000, \ldots\}$.
- $(0 \cup 1)^*1(0 \cup 1)^*$: L contains at least one 1, $L = \{010, 1110, \ldots\}$.
- $(0 \cup 1)^*111$: L ends with three consecutive 1s. For example $L = \{0111, 0110111, \ldots\}$.
- $(00)^*$: L contains even number of 0s. For example, $L = \{00, 0000, 000000, \ldots\}$.
- $(0 \cup 1)^*1(0 \cup 1)(0 \cup 1)$: L contains a 1 as the third symbol from end. For example, $L = \{0111, 01000100, 0101100, \ldots\}$.

Regular expressions and finite state automata are equivalent, in other words, a NFSA describes a regular language. A regular expression is converted to a NFSA

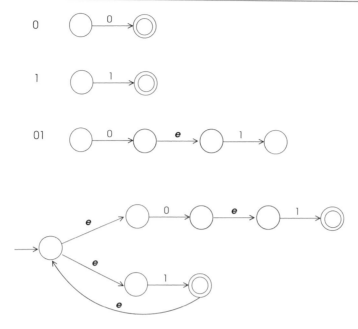

Fig. 10.10 Tree diagram for the NFSA of Fig. 10.8

simply by providing parallel paths for the union operation and cascading the operands for the concatenation operation. The star operation is represented by feedback loops from the accepting states to the beginning states.

Example 10.3 Find the NFSA representation of the regular expression $01 \cup 1^*$ with the alphabet $\sum = \{0, 1\}$.
Solution: We will start from the basic operands of this expression to form the NFSA as shown in Fig. 10.10. First, 0 and 1 transitions are formed and then these are concatenated. The union operation requires forking and the star operation means we need to get back to initial state. Note that transitions labeled by ε are automatically performed.

10.7 Turing Machines

The Turing machine (TM) introduced by Alan Turing is a simplified model of a real computer [3]. It is basically a DFSA with an infinite memory called *tape*. The inputs to TM are the *cells* on the tape and the tape head can read and write a single cell at a time. Each cell contains a symbol from an alphabet and a cell that has not been written before is assumed to contain the blank symbol ⊔.

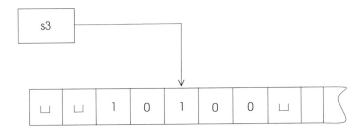

Fig. 10.11 A Turing machine

A TM can both write to and read from the tape and the read-write head can move both to left and right. A TM has *accept* and *reject* states and unlike a FSA, it does not wait until the end of string, a TM simply halts when it enters one of these states. It may never enter one of these states and may run forever in an infinite loop. At each step, a TM,

- Reads the symbol from the cell under the read head,
- Writes a symbol to the tape cell as specified by the transition function δ,
- Moves the read head one cell to the left or to the right, as specified by δ function,
- Changes state as specified by the transition function δ.

A TM initially contains the input string and the empty symbols in any other cell of the tape. A TM at state s_3 with a control unit is depicted in Fig. 10.11. Depending on the current read symbol from the tape, the TM may change its state, move the head left or right and may write to the cell. If the head is in the initial position and needs to move left, it stays in the initial position.

A TM has two alphabets, Σ where all input strings are written in and a tape alphabet Γ where $\Sigma \subseteq \Gamma$. The formal definition of a TM is as follows.

Definition 10.8 (*Turing machine*) A TM is a 7-tuple:

- S: A finite set of states,
- Σ: A finite *input alphabet*,
- Γ: A finite *tape alphabet* where $\sqcup \in \Gamma$ and $\Sigma \subseteq \Gamma$,
- $\delta : S \times \Gamma \to S \times \Gamma \times \{L, R\}$: A transition function,
- $s_0 \in S$: The initial state in S
- $s_{acc} \in S$: The accept state in S
- $s_{rej} \in S$: The reject state in S

A *configuration* of a TM is the union of the input string and the current state. For example,

$$1001s_3\mathbf{0}110$$

represents the configuration C in which tape is 10010110, the current state is s_3 and the head is currently on the symbol 0 shown in bold. Moving from configuration C_x

Fig. 10.12 A Turing
machine basic state
transition

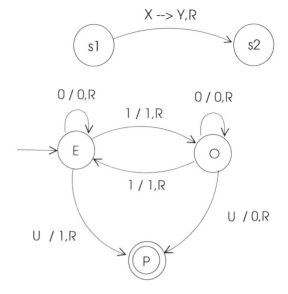

Fig. 10.13 A TM that adds
a bit to a string to have odd
parity

to C_y is achieved by the transition function. A basic state transition of a TM is shown
in Fig. 10.12 where TM reads symbol X under the tape head, replaces it with symbol
Y, changes its state from s_1 to s_2 and moves the head to the right. Changing state is
shown in equation form by $\delta(s_1, X) = (s_2, Y, R)$.

An *accepting configuration* of a TM is a configuration in which the state of this
configuration is the accept state and the state of a *rejecting configuration* is a reject
state. The initial condition of a TM is that the whole string is present in the TM
preceded and followed by infinite blank symbols. A TM accepts an input string if
it enters an accept state, halts if it reaches halt before the end of the string and runs
forever if these two conditions are not encountered.

Example 10.4 Construct a TM that adds a bit to an input string to have odd parity
Solution: The alphabets are $\sum = \{0, 1\}$, $\Gamma = \{0, 1, \sqcup\}$, states are E (even), O (odd)
and P (parity added accepting state) and the initial state is E. The state diagram of
this TM is similar to that of Fig. 10.1 but we need to add a 0 or a 1 to the end of the
string when a $\sqcup$ is encountered as shown in Fig. 10.13. The tape head always moves
to the right and alternates between states E and O when a 1 is read until the blank
symbol is detected. Note that there is no reject state of this TM.

Example 10.5 Design a TM M that accepts the language $L = \{0^n 1^n | n \geq 1\}$.
Solution: We first define the high level description M. It should mark 0s and 1s after
0s but it has to ensure that for every 0 power, there should be a power of 1. The
alphabets are $\sum = \{0, 1\}$ and $\Gamma = \{0, 1, X, Y, \sqcup\}$. The algorithm for M consists of
the following steps:

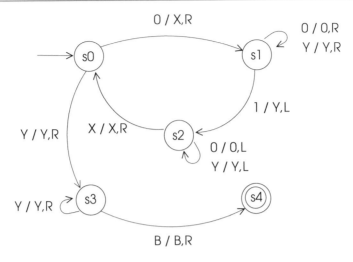

Fig. 10.14 A TM that accepts the language $0^n 1^n$

1. Mark the first unread 0 with X, move right.
2. Move right until the first unread 1 and mark it with Y.
3. Move left until the last marked Y and then move one symbol to the right.
4. If the next position is a 0 then goto 1. Otherwise, move right all along the input string check if if there are unmarked 1s. If not, goto the next blank symbol and accept.

The state diagram of M is depicted in Fig. 10.14 with five states $\{s_0, s_1, s_2, s_3, s_{acc}\}$. State s_3 is a stable state that in which $0^n 1^n$ pairs are detected and when a $\sqcup$ symbol is read at this state, the accept state s_{acc} is entered. Let us check the configuration sequence of M for the input string 0011. The tape contains 0011$\sqcup$ initially and M and he tape head is at the leftmost position under the first 0. The following is the list of configuration transitions of M for this input:

$$s_0 011; \ Xs_1 011; \ X0s_1 11; \ Xs_2 0B1; \ s_2 X0Y1; \ Xs_0 0Y1; \ XXs_1 Y1; \ XXYs_1 1;$$

$$XXs_2 YY; \ Xs_2 XYY; \ XXs_0 YY; \ XXYs_3 Y; \ XXYYs_3 \sqcup; \ XXYY \sqcup s_{acc}\sqcup$$

The state table for M is depicted in Table 10.8.

10.8   Complexity Theory

An *optimization problem* is a problem that we want to obtain the best solution, on the other hand, we provide an instance of a problem and expect an answer as *yes* or *no* in decision problems. Many algorithms we have seen so far have an execution

Table 10.8 State table for M

	0	1	X	Y	⊔
s_0	(s_1, X, R)	–	–	(s_3, Y, R)	–
s_1	$(s_1, 0, R)$	(s_2, Y, L)	–	(s_1, Y, R)	–
s_2	$(s_2, 0, L)$	–	(s_0, X, R)	(s_2, Y, L)	–
s_3	–	–	–	(s_3, Y, R)	$(s_4, ⊔, R)$
s_4	–	–	–	–	–

time expressed by $O(n^k)$ where n is the input size and k is an integer greater than or equal to 0, for example, finding the maximum value of an array is performed in $O(n)$ time. Such algorithms are said to be in the class **P**.

Definition 10.9 P is the set of all decision problems for which there is a polynomial-time algorithm.

A *certificate* of an algorithm is one instance of all possible inputs to a problem. For example, when we want to find the maximum value stored in an integer array, a sample value presented to the algorithm is a certificate. A *certifier* or a *verifier* is an algorithm that tests whether a given certificate provides a *yes* or *no* answer to a given problem. For example, let us assume we need to test whether a given integer m is greater than or equal to all integer values stored in an array. The integer m is the certificate and a simple procedure that runs a *for* loop comparing each value of the array with m is the certifier as shown in Algorithm 10.2. This algorithm runs in $O(n)$ time, thus, it is in **P**.

Algorithm 10.2 *A Certifier*

```
1: procedure CERTIFIER_MAX(m: integer, A[n]:integer)
2:     for i = 1 to n do
3:         if m < A[i] then
4:             return no
5:         end if
6:     end for
7:     return yes
8: end procedure
```

Definition 10.10 (NP) Nondeterministic Polynomial-time (**NP**) is the set of all decision problems that have a polynomial-time certifier.

The problem we have described has a certifier in **P**, thus it is in **NP**. Essentially, all of the problems in **P** have certifiers running in polynomial time and hence we

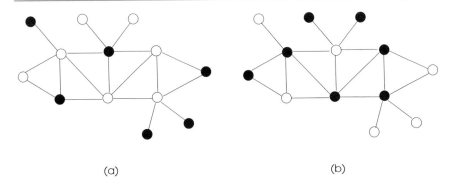

(a)   (b)

Fig. 10.15 a A maximal independents set of a graph G, **b** The vertex cover of the same graph

can deduce **P** ⊆ **NP**. It is not proven until now whether **P** = **NP** but the basic understanding is that this equality does not hold.

10.8.1   Reductions

A computational problem B can be reduced to a problem A in polynomial time in various cases meaning solving problem B is at the same level of difficulty as solving problem A. This relation is shown as $B \leq_p A$ and means if problem A can be solved in polynomial time, problem B can be solved in polynomial time. It also means if A can not be solved in polynomial time, B can not be solved in polynomial time either.

An independent set of a graph $G = (V, E)$ is a set $I \in V$ such that no two vertices in I are adjacent. Let us consider the independent set problem IND which is to find whether a graph G contains an independent set of size greater than or equal to k for $k > 0$. Another question, the vertex cover problem VCOV for a graph $G = (V, E)$ is stated as forming $V' \in V$ with order less than or equal to k for $k > 0$ such that any edge $(u, v) \in E$ has at least one endpoint in V'. We claim these two problems are equivalent and thus need to show $IND \leq_p VCOV$. Let I be an independent with order k of a graph $G = (V, E)$ and $V' = V - I$. There is no edge $(u, v) \in E$ such that $u \in I$ and $v \in I$. Thus, any edge $(u, v) \in E$ must have at least one of its endpoints in V' which means V' is a vertex cover of G.

In the other direction of proof, let $V' = V - I$ be a vertex cover of a graph G. Then any $u, v \in I$ does not have an edge $(u, v) \in E$ between them as otherwise this edge will not be covered by V'. Thus, I is an independent set of G and whenever we say G contains an independent set of size k, this statement is equivalent to saying G contains a vertex cover of order $|V| - k$ as shown in the graph of Fig. 10.15.

Definition 10.11 (*class* **NP**-*Hard*) A decision problem A is **NP**-Hard if every problem in **NP** can be reduced to it in polynomial time. That is, $B \leq_p A, \forall B \in$ **NP**.

Fig. 10.16 Complexity classes

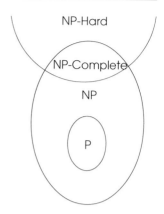

Polynomial time reduction of only one problem in **NP** to a problem A is adequate to show that all problems in **NP** can be reduced to A since polynomial time reduction is transitive.

10.8.2 NP-Completeness

Definition 10.12 A decision problem A is **NP**-Complete if $A \in$ **NP** and every problem in **NP** can be reduced to A in polynomial time.

The relationship between these complexity classes is depicted in Fig. 10.16 where class **P** is a subset of class **NP**. An **NP**-Hard problem A may not be in **NP**, A is as hard as any problem in **NP** and an **NP**-Complete problem is in **NP** and also a member of **NP**-hard problems, thus, it is at the intersection of these two classes.

Basically, we need to perform the following steps to show that a problem A is **NP**-Complete.

- Prove that A is a decision problem,
- Find a certifier that solves an instance of the problem A in polynomial time, hence prove that $A \in$ **NP**,
- Test whether A can be reduced to a problem that is **NP**-Complete.

Let us form a certifier for the IND problem: given a graph $G = (V, E)$ and an integer $k > 0$, does G contain an independent set of at least k vertices? The following algorithm inputs a graph $G = (V, E)$ and a vertex set $I \in V$ of size $k > 0$ and checks whether each edge of the graph has at most one endpoint in I. If both endpoints are contained in I, then it returns *no* as the answer, otherwise a *yes* is returned.

1. **Input**: $G = (V, E)$ and $I \in V$ with $|I| = k, k > 0$
2. **Output**: *yes* or *no*
3. **for all** $u \in I$

4.   **if** $(u, v) \in E$ **then**
5.   **return** *no*
6.   **end if**
7. **end for all**
8. **return** *yes*

The *for* loop is executed $O(n)$ times and checking whether edge (u, v) exists in the graph is done in $O(n)$ times using the adjacency matrix representation resulting in $O(n^2)$ time complexity. Since the algorithm provides a *yes* or *no* answer in polynomial time, it is a certifier. The IND problem can be reduced to VCOV problem as stated and hence we can say IND problem is **NP**-Complete with the assumption that VCOV problem is **NP**-Complete.

10.8.3 Coping with NP-Completeness

When confronted with an **NP**-Complete problem, we can implement one of the following options.

- *Heuristics*: These are common sense rules which do not have proofs. We need to show that a heuristic H works with favorable performance across a wide range of inputs to consider it as a good heuristic. For example, the vertex coloring problem was selecting minimal number of colors to assign to the vertices of a graph such that no two adjacent vertices receive the same color as reviewed in Chap. 13. A seemingly reasonable heuristic to color a graph could be starting coloring from the highest degree vertex and always selecting the next highest degree vertex to color at each step. However, this heuristic does not always give even suboptimal results.
- *Approximation Algorithms*: An approximation algorithm finds a suboptimal solution to a given problem in polynomial time. We will review such an algorithm in Chap. 13 that finds the vertex cover of a graph by finding a maximal matching and including endpoints of matched edges in vertex cover with an approximation ratio of 2, that is, it finds a vertex cover of order which is twice the order of the minimum vertex cover. Unlike using heuristics, an approximation algorithm is proven to always work with the approximation ratio.
- *Algorithmic Approaches*: Backtracking through the search space of an algorithm eliminates some of the possibilities and thus may be used for some difficult problems. In this method, all possible states of an algorithm is formed as a state-space tree. Branches of the tree are traversed and whenever a requirement is violated, return to a previous state is provided. Thus, extensive search of all possibilities is prevented. Another approach is to use a randomized algorithm in which the execution pattern of the algorithm depends on some randomly generated numbers or random choices. A *Las Vegas algorithm* always returns correct answer with random time and a *Monte Carlo algorithm* runs in constant time but may return a wrong answer.

10.9   Review Questions

1. What is an alphabet and a language defined on an alphabet?
2. Compare state diagram and state table representation of a finite state machine. Which model is better suited for programming?
3. What are the main differences between the finite state machine and the finite state automata models?
4. Describe how a finite state automata recognizes a language.
5. Compare deterministic finite state automata with nondeterministic finite state automata.
6. What is a regular expression and how is it related to a language?
7. What is the relationship between a regular language and a finite state automata?
8. What makes a Turing Machine different than a finite state automata?
9. What type of application can a Turing machine be used?
10. What is the complexity class **P**?
11. How can we determine whether a problem is in **NP**?
12. What is meant by reducing a problem to another problem?
13. What are the main test steps to determine whether a problem is **NP**-Complete?

10.10   Chapter Notes

Theory of computation has complexity theory, computability theory, and automata theory as the main components. We reviewed mainly automata theory in this chapter with emphasis on languages and finite state automata. A language is basically a string of symbols over some alphabet. A finite state machine is used to model a system where the output of the machine depends on the current input and the current sate. A finite state automaton is a state machine with no outputs but with accepting states. An automaton is used to recognize a language, if it reaches an accepting state at the end of input, then the input language is recognized. A nondeterministic finite state automaton may take empty string as an input and it may divert to more than one state with the same input. Whenever such diversion occurs, a copy of the automaton is run in parallel and if one of the running branches of the automaton reaches an accepting state, the input is accepted and computation stops. A regular expression describes a language in short form. A Turing machine is more general than an automaton as it may accept input strings that may not be accepted by the automaton. It consists of a control unit, an infinite tape with symbols and read/write tape head. The head may move left or right and each move may result in a state transition of the machine. Turing machines represent a computer closely and may be used for a wide range of applications.

In the last part of the chapter, we reviewed the basic complexity classes **P** , **NP**, **NP**-Hard and **NP**-Complete. Any problem that can be solved by a polynomial time algorithm is in class **P** and a problem that has a certifier which runs in polynomial time is in **NP**. The class **NP**-Hard denotes problems that can be reduced to problems in

NP in polynomial time and a problem that is in **NP** and is **NP**-Hard is **NP**-Complete. In order to show whether a problem A is as hard as another problem B, reduction of A to B is commonly performed. We have barely touched the surface of the theory of computation in this chapter. A fine and detailed study of the theory of computation is given in [1] and a practical textbook is provided in [2].

Exercises

1. Convert the FSM of a odd parity checker to a FSA and show the transitions for the binary input string 100101101.
2. Design a DFSA M which detects 101 as a substring in any given binary input string with the alphabet $\sum = \{0, 1\}$.
3. Design a NFSA M which accepts the string $aabbab$ with the alphabet $\sum = \{a, b\}$.
4. State the languages described by the following regular expressions by giving examples.

   a. $0 \cup 1)10(0 \cup 1)$ c. $(111)^*$
   b. 10^* d. $(0 \cup 1)^* 11(0 \cup 1)^*$

5. Design a NFSA M which represents the regular expression $(0 \cup 1)^* 10$ with the alphabet $\sum = \{0, 1\}$.
6. Provide the pseudocode of a certifier for the dominating set problem DOM: Is there a dominating set with order less than or equal to k in a graph G?

References

1. Sipser M (2013) Introduction to the theory of computation, 3rd edn. CENGAGE Learning. ISBN-10: 1-133-18781-1
2. Maheshwari A, Smid M (2019) Introduction to theory of computation. Free textbook. Carleton University, Ottawa
3. Turing A (1937) On computable numbers, with an application to the Entscheidungs problem. In: Proceedings of the London mathematical society, Series 2, vol 42, pp 230–265. https://doi. org/10.1112/plms/s2-42.1.230

Graph Theory

Part II

Introduction to Graphs

11

Graphs are key data structures that find numerous applications. A graph has a number of nodes some of which are connected. This simple structure can be used to represent many real-life applications such as a road network, communication network and a biological network. In fact, any network may be modeled by a graph and the methods of graph theory can be implemented conveniently to solve various problems in these networks. We define graphs, review types, operations on graphs and graph representations in this chapter to form the basic background for further chapters in this part.

11.1 Terminology

A graph consists of a set of points and a set of lines joining some of these points. The points are the *vertices*, or *nodes* as commonly called, of the graph and the lines are the *edges* between the vertices.

Definition 11.1 (*Graph*) A graph $G = (V, E)$ is a discrete structure consisting of a vertex set V and an edge set E.

The vertex set of a graph G is denoted by $V(G)$, and its edge set by $E(G)$. When the graph under consideration is clear, we omit G and use V and E for vertex and edge sets of graph G. The *order* of a graph G is the number of its vertices and its *size* is the number of its edges. We will use literal n to denote the order and m to denote the size of a graph. Commonly, vertices of a graph have labels as literals such as a, b, c etc. or integers such as 1, 2, 3. An edge is shown as (a, b) or ab or $\{a, b\}$ where a and b are the vertices the edge is connected. The vertices a and b are called the *endpoints* of edge (a, b). An example graph is depicted in Fig. 11.1 with

© Springer Nature Switzerland AG 2021
K. Erciyes, *Discrete Mathematics and Graph Theory*, Undergraduate Topics in Computer Science, https://doi.org/10.1007/978-3-030-61115-6_11

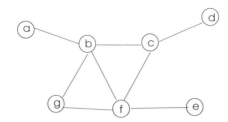

Fig. 11.1 A sample graph

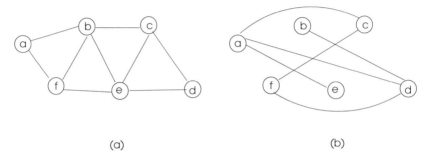

(a)   (b)

Fig. 11.2 a A sample graph **b** Its complement

$V = \{a, b, c, d, e, f, g\}$ and $E = \{(a, b), (b, c), (c, d), (c, f)(b, f), (f, g), (b, g)\}$ and the values of n and m are 7 and 8 respectively.

A graph may contain *multiple edges* which are edges with the same endpoints. A *self-loop* starts and end at the same vertex. An edge that is not a self-loop is called a *proper edge*. A *simple graph* does not contain any self-loops or *multiple edges*. A graph with multiple edges is called a *multigraph* and a *finite graph* has a finite number of vertices and finite number of edges. We will consider mainly simple and finite graphs in this text due to their numerous implementations in practice. A multigraph may be converted to its *underlying* graph by substituting a single edge for each multiedge. An *undirected graph* has no orientation of its edges.

Definition 11.2 (*Complement of a graph*) The *complement* of a graph $G = (V, E)$ is the graph $G' = (V, E')$ such that an edge $(u, v) \in E'$ if and only if $(u, v) \notin E$.

Basically, G' contains the same vertex set as G but an edge contained in G is not included in G'. Complement of a sample graph is shown in Fig. 11.2.

11.2 Vertex Degree

Definition 11.3 (*Degree of a vertex*) The number of of edges incident to a vertex v in an undirected graph G is called the *degree* of v and is denoted by $deg(v)$.

Self-loops of a vertex are counted two to find the degree of the vertex. The degrees of vertices in the graph of Fig. 11.1 for vertices $a, b, ..., g$ are 1, 4, 3, 1, 1, 4, 2 respectively. The maximum degree in a graph G is denoted by $\Delta(G)$ or simply by Δ when the graph under consideration is clear. The minimum degree in a graph G is denoted by $\delta(G)$ or simply by δ. The degree of a vertex v in a graph G, $deg(v)$, is between the maximum and minimum degree of G as shown below.

$$0 \leq \delta(G) \leq deg(v) \leq \Delta(G) \leq n - 1 \tag{11.1}$$

A vertex with an odd degree is called an *odd vertex* and a vertex with an even degree is called an *even vertex*. A vertex that has a degree of 0 is called An *isolated vertex* and a *pendant vertex* has a degree of 1. The vertices a, d and e are pendant vertices in the graph of Fig. 11.1.

Remark 11.1 The sum of the degrees of vertices in a graph is an even number. Let us start with a graph consisting of a single vertex which has a total 0 degree. Every time an edge is added to this graph, the sum of degrees increases by 2.

Theorem 11.1 (Euler) *The sum of the degrees of a simple undirected graph $G = (V, E)$ is twice the number of its edges. Formally,*

$$\sum_{v \in V} deg(v) = 2m \tag{11.2}$$

An edge (u, v) is counted twice, once for being incident to vertex a and once for vertex b, hence total degree of the graph is twice the number of edges. Number of edges, m, of the graph in Fig. 11.1 is 8 and the sum of the degrees is 16. The average degree of an undirected graph is the arithmetic average of all of the degrees of vertices as below,

$$\frac{\sum_{v \in V} deg(v)}{n} = 2m/n \tag{11.3}$$

The average degree of the graph of Fig. 11.1 is $16/7 \approx 2.3$.

Corollary 11.1 *There is an even number of odd vertices in an undirected graph.*

Proof Let the set of even vertices of a graph G be V_E and the set of odd vertices be V_O. Then, the following equation may be stated,

$$\sum_{v \in V} deg(v) = \sum_{v \in V_O} deg(v) + \sum_{v \in V_E} deg(v)$$

Since the sum of the degrees of G is $2m$ by Theorem 11.1, the left side of this equation is an even number. Let the sum of degrees of odd vertices be S_O and the sum of degrees of even vertices be S_E which means $2m = S_O + S_E$ by the above equation. The sum S_E is even as it consists of the sum of all even numbers, thus, S_O must also be even to result in the even total sum of $2m$. Since S_O is even, it must be

the sum of even number of odd terms as the sum of odd number of odd numbers is
odd.   □

Theorem 11.2 *Given an undirected graph* $G = (V, E)$ *with* $n \geq 2$, *there exists*
$u, v \in V$ *such that* $deg(u) = deg(v)$.

Proof We will use contradiction to prove this theorem. Let us assume the opposite
that a graph with n vertices does not have a single vertex pair u and v such that
$deg(u) = deg(v)$. All vertices have distinct degrees in this case and the only possible
degree values are $0, ..., (n-1)$. A vertex u with a degree of 0 means u is an isolated
vertex and a vertex v with a degree of $n-1$ means v is connected to all other vertices
of the graph which means u can not be an isolated vertex, thus a contradiction.   □

   As an implementation of this result, consider a computer network consisting of
routers connected to other routers. Based on Theorem 11.2, there are at least two
routers in any such computer network that are connected to the same number of
routers.

11.2.1 Degree Sequence

The degree sequence of a graph $G = (V, E)$ is a monotonic non-increasing sequence
of the degrees of its vertices. The degree sequence of the graph in Fig. 11.1 is
$\{1, 1, 1, 2, 3, 4, 4\}$. If a degree sequence $D = (d_1, d_2, ..., d_n)$ represents a degree
sequence of some graph G, D is called *graphical*. Note that some degree sequences
may not represent a realizable graph. For example, the sequence $\{1, 2, 4, 5, 5\}$ can
not represent a real graph as the sum is an odd number and we know by Theorem 11.1
the sum of degrees of a graph is always an even number. We can have different graphs
with the same degree sequence.

11.3 Directed Graph

In an undirected graph with no direction of edges, and edge between the vertices
a and b is denoted by (a, b) or (b, a) as noted. The edges of a graph may be oriented in which case an arrow indicates the direction. Such graphs are called *directed
graphs* or *digraphs* and an edge (a, b) in such a graph indicates this edge is directed from vertex a to b which is shown by $(a, b) \in E$ but it does not mean means
$(b, a) \in E$. A *partially directed graph* has has both directed and undirected edges.
A *complete simple digraph* has a pair of edges, one in each direction, between any
two vertices it has. A digraph is depicted in Fig. 11.3 with $V = \{a, b, c, d, e\}$ and
$E = \{(a, b), (a, f), (b, c), (b, f), (d, c), (e, c), (e, d), (f, e), \}$.

Fig. 11.3 A sample digraph

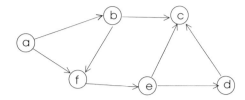

Definition 11.4 (*Indegree*) The indegree $d_{in}(v)$ of a vertex v in a digraph $G = (V, E)$ is the number of edges oriented towards v Formally,

$$d_{in}(v) = |\{(u, v) \in E\}|$$

Definition 11.5 (*Outdegree*) The outdegree $d_{out}(v)$ of v is the number of edges coming out from this vertex. Formally,

$$d_{out}(v) = |\{(v, u) \in E\}|$$

The degree of a vertex in a digraph is the sum of its indegree and its outdegree. The sum of the indegrees of the vertices in a graph is equal to the sum of the outdegrees which are both equal to the sum of the number of edges as shown below.

$$\sum_{v \in V} d_{in}(v) = \sum_{v \in V} d_{out}(v) = m$$

Informally, this result is due to each directed edge increasing total indegree value by 1 and total outdegree by 1.

11.4 Representation of a Graph

We have three basic representations of a graph as the adjacency list, adjacency matrix and incidence matrix representations.

11.4.1 Adjacency List

In *adjacency list representation* of a simple graph or a digraph, a vertex v of the graph is represented as a linked list with the head of the list being vertex v as shown in Fig. 11.4. The end of the list is signified by the last node having a *null* pointer which means a pointer with an empty value showing nothing. All of the neighbors of a vertex v are linked in the linked list of vertex v. This way of representation is commonly used for sparse graphs where $m \ll n$. A graph can be stored in $O(n+m)$ space, n for list heads and 2m for edges, and searching an edge can be performed in $O(n)$ time since the maximum length of the linked list for a node can have $n - 1$ elements.

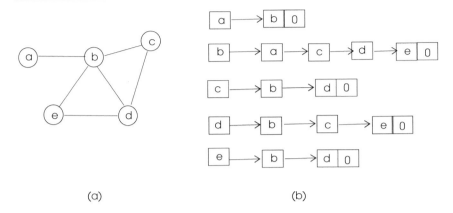

(a)   (b)

Fig. 11.4 **a** A sample graph, **b** Its adjacency list

Fig. 11.5 A digraph

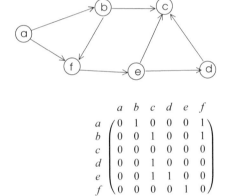

Fig. 11.6 Its adjacency
matrix

$$
\begin{array}{c c c c c c c}
 & a & b & c & d & e & f \\
a & 0 & 1 & 0 & 0 & 0 & 1 \\
b & 0 & 0 & 1 & 0 & 0 & 1 \\
c & 0 & 0 & 0 & 0 & 0 & 0 \\
d & 0 & 0 & 1 & 0 & 0 & 0 \\
e & 0 & 0 & 1 & 1 & 0 & 0 \\
f & 0 & 0 & 0 & 0 & 1 & 0 \\
\end{array}
$$

11.4.2   Adjacency Matrix

An adjacency matrix $A[n, n]$ of a graph $G = (V, E)$ has $A[i, j] = 1$ if there is
an edge between the vertices i and j in the graph and $A[i, j] = 0$ if no such edge
exists. Clearly, matrix A is symmetric, that is, $A[i, j] = A[j, i]$ for all $i, j \in E$ if G
is undirected. The adjacency matrix A will not be symmetric for digraph in general
since an edge $(a, b) \in E$ does not imply $(b, a) \in E$. The entry $A[i, j]$ may be greater
than 1 showing the number of edges between vertices i and j in a multigraph. The
matrix A for the graph of Fig. 11.5 is shown in Fig. 11.6.

Determining whether and edge (a, b) exists in a graph G can be done in one step
by checking matrix entry $A[a, b]$. Labeling vertices with positive integer sequence is
reasonable to be able to perform this matrix search quickly. However, space required
for the adjacency matrix is $O(n^2)$ which may be significant for a large graph.

Fig. 11.7 A digraph

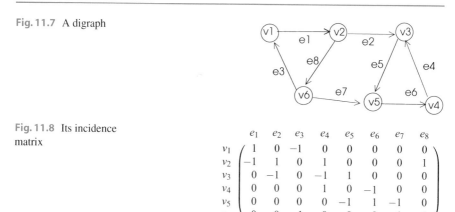

Fig. 11.8 Its incidence matrix

$$
\begin{array}{c c c c c c c c c}
 & e_1 & e_2 & e_3 & e_4 & e_5 & e_6 & e_7 & e_8 \\
v_1 & 1 & 0 & -1 & 0 & 0 & 0 & 0 & 0 \\
v_2 & -1 & 1 & 0 & 1 & 0 & 0 & 0 & 1 \\
v_3 & 0 & -1 & 0 & -1 & 1 & 0 & 0 & 0 \\
v_4 & 0 & 0 & 0 & 1 & 0 & -1 & 0 & 0 \\
v_5 & 0 & 0 & 0 & 0 & -1 & 1 & -1 & 0 \\
v_6 & 0 & 0 & 1 & 0 & 0 & 0 & 1 & -1
\end{array}
$$

11.4.3 Incidence Matrix

An incidence matrix B of an undirected graph $G = (V, E)$ has $n \times m$ elements and $B[i, j] = 1$ if edge e_j is incident to vertex v_i in G and $B[i, j] = 0$ if edge j is not incident to that vertex. Note that this is a relation between the vertices and edges of a graph and we need to label edges now to be able to form the matrix B. The incidence matrix of a directed graph is again an $n \times m$ matrix B such that $B[i, j] = 1$ if the edge e_j originates from vertex v_i, -1 if it ends in vertex v_i and 0 otherwise. An example graph and its incidence matrix is depicted in Figs. 11.7 and 11.8.

11.5 Subgraphs

In many cases, the focus of analysis may be confined to a certain part of a graph rather than the whole graph. This is indeed the case when graph under consideration is very large, consisting of thousands of vertices and tens of thousands of edges. A subgraph of a graph G consists of some vertices and/or some edges of G. Some graph applications require searching a subgraph with a specific property.

Definition 11.6 (*Subgraph*) A graph $G' = (V', E')$ is called a *subgraph* of graph $G = (V, E)$ if $V' \subseteq V$ and $E' \subseteq E$ and G is called a *supergraph* of G'. If the subgraph G' of G is not equal to G, then G' is called a *proper subgraph* of G.

A subgraph $G' = (V', E')$ of a graph $G = (V, E)$ is called an *induced subgraph* of G if $\forall u, v \in V'$, $(u, v) \in E'$ if and only if $(u, v) \in E$. In other words, if two vertices contained in the induced subgraph G' are connected in G, then they are also connected in G'. A *spanning subgraph* G' of a graph G contains all vertices of G. A graph, its proper subgraph and induced subgraph are depicted in Fig. 11.9.

Deleting a vertex v from a graph G is done by deleting v and all of its incident edges from G. The obtained subgraph as the result of this operation is denoted by $G - v$. Subgraph obtained after deletion of a set V' of vertices from G is shown by

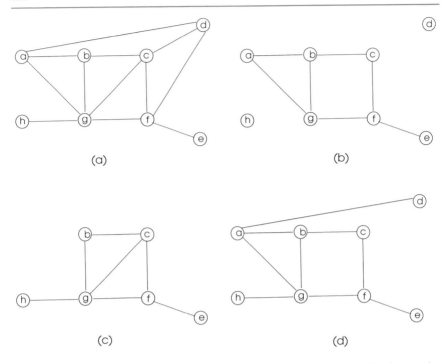

Fig. 11.9 **a** A sample graph G, **b** a proper subgraph of G, **c** an induced subgraph of G, **d** a spanning subgraph of G

$G \setminus V'$ or $G - V'$ and the induced subgraph of G when V' is deleted is denoted by $G[V']$. Similarly, an edge e may be deleted from a graph G, and the resulting subgraph is shown by $G - e$. Note that deleting an edge does not remove any vertices from the graph.

11.6   Types of Graphs

Certain graph types are interesting to analyze as they commonly represent real-life situations. The main graph types are complete graphs, weighted graphs and bipartite graphs as described next.

11.6.1   Complete Graph

Every vertex is connected to all other vertices in a *complete graph*. Such graphs are shown by K_n where n is the number of vertices. Complete graphs $K_1, .., K_6$ are depicted in Fig. 11.10. A complete digraph has directed edges in both directions for each vertex pair it has.

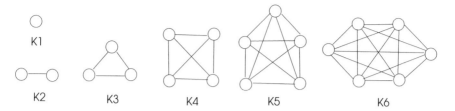

Fig. 11.10 Complete graphs of orders 1 to 6

The number of edges of a simple undirected complete graph K_n can ce calculated as follows. The degree of each vertex in K_n is $n - 1$ and there are n vertices meaning the sum of degrees is $n(n - 1)$. Since two endpoints of an edge are counted to find the sum of degrees, we need to divide the sum by 2 to get the number of edges as $n(n - 1)/2$.

11.6.2 Weighted Graphs

An edge-weighted graph has real numbers associated with its edges. These numbers may represent cost of going from one node to another as in the case of a road network represented by a graph, or the cost of sending a packet from one computer network node to another when a graph represents such a network.

Definition 11.7 (*Edge-weighted graphs*) The edges of an edge-weighted graph $G(V, E, w)$, $w : E \rightarrow \mathbb{R}$ are labeled with real numbers.

The vertices of a graph may have weights representing a function on that node, for example memory capacity of a node in a computer network.

Definition 11.8 (*Vertex-weighted graphs*) A vertex-weighted graph $G(V, E, w)$, $w : V \rightarrow \mathbb{R}$ has vertices labeled with real numbers.

11.6.3 Bipartite Graphs

The vertex set of a graph may be divided into two distinct sets such that there are no edges between any vertices in each set. Such graphs are called bipartite graphs.

Definition 11.9 (*Bipartite graph*) A *bipartite graph* $G = (V, E)$ or $G = (V_1 \cup V_2, E)$ has two distinct vertex sets V_1 and V_2 such that $\forall (u, v) \in E$, either $u \in V_1$ and $v \in V_2$, or $u \in V_1$ and $v \in V_2$.

Fig. 11.11 A sample
bipartite graph with
$V_1 = \{a, b, c\}$ and
$V_2 = \{d, e, f, g, h\}$

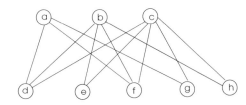

In other words, any edge of a bipartite graph joins a vertex from one vertex set to the other. A bipartite graph is depicted in Fig. 11.11. A directed bipartite graph has edges with orientation and a weighted bipartite graph has weights associated with its edges. A complete bipartite has edges between each vertex of the first set of vertices to all other vertices of the second vertex set. Such a graph is denoted by $K_{m,n}$ where m and n are the number of vertices in the disjoint vertex sets. A complete bipartite graph $K_{m,n}$ contains mn edges and $m + n$ vertices.

11.6.4 Regular Graphs

A regular graph has vertices with the same degree. When the common degree is k, the graph is called a k-regular graph. Figure 11.12 shows 0 to 4-regular graphs. Note that the 3-regular graph is K_4 and 4-regular graph is K_5. In general, every K_n graph is a $(n-1)$-regular graph but not every $(n-1)$-regular graph is a K_n graph. For example, 2-regular graph shown in Fig. 11.12b is not a complete graph.

11.6.5 Line Graphs

A line graph L of a graph G is obtained by representing each edge of G as a vertex in L and connecting two vertices in L only if they have common endpoints in G. Many properties of a graph are translated to its line graph. A graph and its line graph are shown in Fig. 11.13.

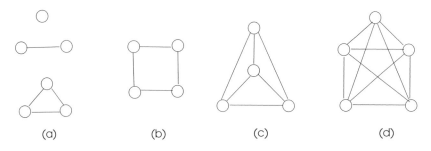

(a)   (b)   (c)   (d)

Fig. 11.12 **a** 0-regular, 1-regular and 2-regular graphs **b** a 3-regular graph **c** a 4-regular graph

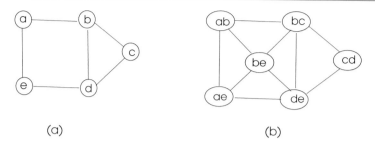

Fig. 11.13 **a** A sample graph G, **b** Its line graph

The line graph L of a graph $G = (V, E)$ has $n' = m$ nodes and m' edges as follows [7],

$$m' = \frac{1}{2} \sum_{v \in V} deg(v)^2 - m$$

For the example in Fig. 11.13,

$$m' = deg(a)^2 + deg(b)^2 + deg(c)^2 + deg(d)^2 + deg(e)^2$$
$$= (4 + 9 + 4 + 9 + 4)/2 - 6 = 9$$

The incidence matrix B of a graph G and the adjacency matrix $A(L)$ of its line graph L are related as follows [7],

$$A(L) = B^T B - 2I$$

11.7 Graph Operations

A graph operation takes at least two graphs as arguments and forms a new graph as result of the operation. Common graph operations are union, intersection and cartesian product.

11.7.1 Graph Union

The union of two graphs has all vertices and all edges of the two graphs after this operation. A formal definition is as follows.

Definition 11.10 (*Union of two graphs*) The *union* of two graphs $G = (V_G, E_G)$ and $H = (V_H, E_H)$ is a graph $F = (V_F, E_F)$ in which $V_F = V_G \cup V_H$ and $E_F = E_G \cup E_H$.

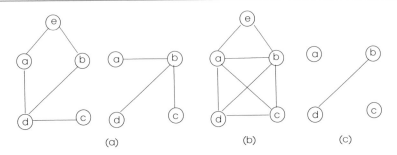

Fig. 11.14 **a** Two input graphs, **b** their union **c** their intersection

When two graphs G and H are distinct, then their union F is a graph with F and G as disjoint subgraphs. The union of two graphs with common vertices and edges is depicted in Fig. 11.14b.

11.7.2 Graph Intersection

The intersection of two graphs has only common vertices and common edges of the two graphs after this operation. A formal definition is given below.

Definition 11.11 (*Intersection of two graphs*) The *intersection* of two graphs $G = (V_G, E_G)$ and $H = (V_H, E_H)$ is a graph $F = (V_F, E_F)$ in which $V_F = V_G \cap V_H$ and $E_F = E_G \cap E_H$.

The intersection of two graphs with common vertices and edges is depicted in Fig. 11.14c.

11.7.3 Graph Join

Graph join is different than the union as this operation creates new edges between each vertex pairs of the two input graphs. A formal definition is given below.

Definition 11.12 (*Join of two graphs*) The *join* of two graphs $G = (V_G, E_G)$ and $H = (V_H, E_H)$ is a graph $F = (V_F, E_F)$ in which $V_F = V_G \cup V_H$ and $E_F = E_G \cup E_H \cup \{(u, v) : u \in V_G \text{ and } v \in V_H\}$. This operation is shown as $F = G \vee H$.

That is, every vertex of G is connected with every vertex of H when F is formed by also keeping the existing edges in both input graphs. When graphs G and H have common vertices, self-loops and multiple edges are formed and these are discarded in the resulting graph. Two input graphs with common vertices a and b are displayed in Fig. 11.15a, their first join formed and simplified join are shown in (b) and (c). The

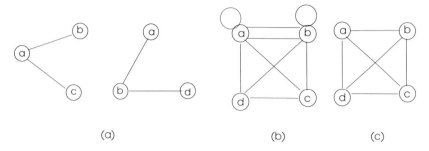

Fig. 11.15 Graph join operation

union, intersection and join operations are commutative, such that $G \cup H = H \cup G$, $G \cap H = H \cap G$ and $G \vee H = H \vee G$.

11.7.4 Cartesian Product

As we formed cartesian product of two sets, we can form the cartesian product of two graphs, this time, forming product of vertices and edges.

Definition 11.13 (*Cartesian product*) The cartesian product of two graphs $G = (V_G, E_G)$ and $H = (V_H, E_H)$ shown by $G \square H$ or $G \times H$ is a graph $F = (V_F, E_F)$ in which $V_F = V_G \times V_H$ and two vertices (u, u') and (v, v') are adjacent in F if and only if one of the following conditions holds:

1. $u = v$ and $(u', v') \in E_H$.
2. $u' = v'$ and $(u, v) \in E_G$.

Thus, each vertex in the product graph F represents two vertices u and v, with $u \in V_G$ and $v \in V_H$. The edges in F are formed using the above rules. The cartesian

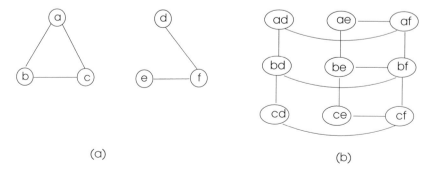

Fig. 11.16 Cartesian product of two graphs

product of two graphs is displayed in Fig. 11.16. For example, edge (ad, af) is included in the product using the first rule and edge (bd, cd) is included using the second rule.

11.8 Connectivity

We are often interested whether we can follow a path from a given vertex to another exists in a given graph. This property needs to be maintained in various type of networks. First, let us review some basic definitions.

11.8.1 Definitions

Definition 11.14 (*Walk*) A *walk* between two vertices v_x and v_y of a graph G is an alternating sequence of $n+1$ vertices and n edges as $W = (v_x, e_1, v_p, e_2, ..., v_q, e_n, v_y)$ where e_i is incident to vertices v_{i-1} and v_i. The vertex v_x is called the *initial* vertex and v_y is called the *terminating* vertex of the walk W.

A walk in the graph of Fig. 11.17 shown by bold edges is $W = \{v_{11}, e_{13}, v_{10}, e_{12}, v_9, e_6, v_2, e_4, v_4, e_5, v_9, e_{11}, v_8\}$ with v_{11} as the initial vertex and v_8 as the terminating vertex. The length of a walk is the number of edges contained in it, the length of W in the above example is 6. A walk may go through the same vertex and edge more than once as in the example walk W where v_9 is visited twice. Lastly, a *closed walk* starts and ends at the same vertex and an *open walk* has different initial end terminating vertices. What we have said up to this point is valid for digraphs.

We need few more definitions before we can review the connectivity concept in graphs. A *trail* is a walk with no repeated edges and a *path* shown by vertices only is a trail with no repeated vertices. The walk W in our example is a trail but is not a path since vertex v_9 is visited twice. The dashed edges represent a path $P = \{v_1, v_2, v_3, v_4, v_8, v_7\}$ in Fig. 11.17. A path with the same initial and terminating vertex is called a *cycle*. The path $P = \{v_2, v_4, v_8, v_9, v_2\}$ is a cycle in this figure. A trail that starts and ends at the same vertex is called a *circuit*. Note that a trail allows

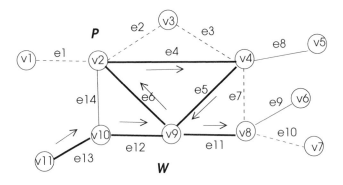

Fig. 11.17 A walk and a path in a sample graph

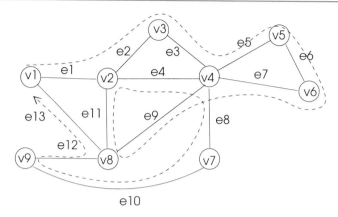

Fig. 11.18 A Eularian graph

repetition of vertices, thus, a circuit may have repeated vertices but not repeated edges. The *girth* of a graph G is the length of the shortest cycle contained in G.

Definition 11.15 (*Eularian tour, Eularian cycle*) An *Eularian tour* of a graph G is a closed trail that visits each edge exactly once and an *Eularian graph* is a graph that has an Eularian tour. An Eulerian cycle is an Eularian tour that starts and ends at the same vertex.

In other words, edges in an Eularian tour can not be repeated but a vertex may be visited more than once. Figure 11.18 displays an Eularian graph with an Eularian cycle that visits edges $\{e_1, e_2, e_3, e_5, e_6, e_7, e_9, e_{11}, e_4, e_8, e_{10}, e_{12}, e_{13}\}$ as shown by the dashed curve.

Definition 11.16 (*Hamiltonian Cycle*) A cycle that visits each vertex of a graph exactly once is called a *Hamiltonian cycle* and such a graph is called *Hamiltonian*.

A Hamiltonian graph with a Hamiltonian cycle shown using a dashed curve that visits vertices $v_1, v_3, v_4, v_5, v_6, v_2, v_7, v_8, v_1$ in sequence is depicted in Fig. 11.19. The number of edges contained in a cycle is denoted as its length l and shown as C_l. For example, C_3 is a triangle.

11.8.2 Connectedness

There is a path between every pair of vertices in a *connected graph*. A *disconnected* graph has subgraphs called *components* that are not connected to each other. A connected graph has one component. Connectivity for a digraph needs to be considered

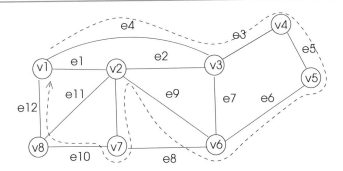

Fig. 11.19 A Hamiltonian graph

carefully. A digraph G is connected if the underlying graph of G is connected. A strongly connected digraph has paths between each pair of vertices in both directions.

Definition 11.17 (*Distance*) Let u and v be any two vertices of an unweighted undirected graph or a digraph G. The distance $d(u, v)$ between u and v is the length of the shortest path between them which is equal to the number of edges in this path.

In a weighted graph, distance between vertices u and v, $d(u, v)$, is the sum of the weights of the edges of the path that joins these vertices. The shortest path between vertices v_4 and v_5 is shown in dashed line in Fig. 11.20 where $d(v_4, v_5) = 11$.

Definition 11.18 (*Eccentricity*) The *eccentricity* of a vertex u in a connected graph $G = (V, E)$ is defined below,

$$max(d(u, v)), \quad \forall v \in V$$

Thus, eccentricity of a vertex u is its maximum distance to any other vertex in the graph. The maximum value of eccentricity in a graph G is called the *diameter* of G and the minimum eccentricity value out of all vertex eccentricities is the *radius* of G. The diameter of the graph in Fig. 11.20 is 4 and its radius is 2 since vertices v_2 and v_7 have both eccentricities of 2 and this value is the minimum value out of all eccentricities in G. Vertex or vertices of a graph G that have the minimum

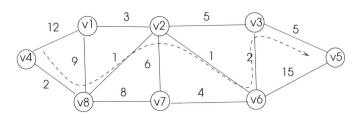

Fig. 11.20 Shortest path between two vertices in a weighted graph

eccentricity value are called the center(s) of G. In the example graph, vertices v_2 and v_7 are the centers of the graph. Note that such vertices should be within the central area of a graph to have low eccentricity values. The diameter of a graph is an important parameter in any network that shows the time it takes to transfer information between two farthest points in such a network. The length of a cycle in a graph can be an odd integer in which case the cycle is called an *odd cycle*, otherwise, it is called an *even-cycle*.

11.9 Graph Isomorphism

Two graphs may be the same except the naming of their vertices. Informally, two graphs G and H are isomorphic to each other if their vertices are labeled different.

Definition 11.19 (*Graph isomorphism*) Let G and H be two simple graphs. A graph isomorphism from $G = (V_G, E_G)$ to $H = (V_H, E_H)$ is a bijection $\phi : V_1 \rightarrow V_2$ with the following condition:

$$(u, v) \in E_G \leftrightarrow (\phi(u), \phi(v)) \in E_H$$

When this condition holds, G and H are said to be isomorphic denoted by $G \approx H$. An isomorphism of a graph to itself is called an *automorphism*.

Two isomorphic graphs may not look similar as shown in Fig. 11.21a and b where a lowercase letter vertex is mapped to an uppercase letter in both examples.

Graph isomorphism problem is to determine whether two given graphs are isomorphic to each other. There is no known polynomial time algorithm to solve this problem. Let us consider two graphs G and H each with n vertices and m edges. Using a naive approach, we can check each one-to-one correspondences between vertices in $n!$ time and edge correspondences in $m!$ time for a total of $n! \cdot m!$ time which grows very fast. We can however check whether two graphs are not isomorphic in simpler ways. For example, if two graphs have different number of vertices, we can conclude they are not isomorphic due to the required one-to-one relation between the vertices.

A property that is preserved when two graphs G and H are isomorphic is called an *isomorphic invariant*. In other words, when graph G has this property, then H has the same property. The list of main isomorphic invariants are listed in the following.

- Having n vertices.
- Having m edges.
- Having the same number of vertices of degree k.
- Being connected.
- Having an Euler circuit.
- Having an Hamiltonian circuit.

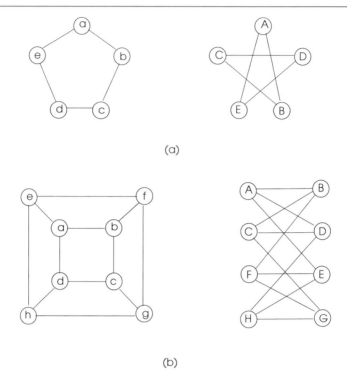

(a)

(b)

Fig. 11.21 Isomorphic graph examples

11.10   Review Questions

1. What is meant by the degree of a vertex in an undirected graph and a digraph?
2. How is the complement of a graph obtained?
3. What is a graphical degree sequence?
4. What are the main methods of representing graphs?
5. What is a subgraph and a proper subgraph of a graph?
6. What is a weighted graph?
7. How is a bipartite graph defined?
8. What is a regular graph?
9. How is a line graph of a graph obtained?
10. What is the resulting graph after the union operation of a graph and its complement?
11. Compare graph union and graph join operations.
12. What is the difference between a trail and a path in a graph?
13. What is an Eularian tour and an Eularian cycle?
14. Does every graph have an Eulerian tour?
15. What is an Hamiltonian tour and an Hamiltonian cycle?
16. What is the distance between two vertices in an unweighted and weighted graph?
17. What is the eccentricity and the radius of a graph?
18. What is meant by two graphs G and H being isomorphic to each other?

11.11 Chapter Notes

This chapter serves as an introduction to the main concepts in graph theory. We started with basic definitions of a graph, degree of a vertex and then described digraphs which have orientation in their edges. Graphs may be represented by adjacency lists, adjacency matrices and incidence matrices. A proper subgraph of a graph contains some of its vertices and/or some of its edges. Weighted, bipartite and line graphs are commonly encountered in practice when a graph is used in a real application.

Basic graph operations are union, intersection, join and cartesian product of two or more graphs. We are often interested in finding whether it is possible to follow a path which is a sequence of vertices between any vertex pairs in a graph. Walks, trails, paths, cycles comprise the terminology to analyze how to reach from one vertex to another in a graph. An Eulerian tour of a graph visits each edge exactly once and a Hamiltonian path of a graph visits each vertex of the graph exactly once. Two graphs are isomorphic to each other if their vertices have different labels. A graph is called a tree if it is connected and does not contain any cycles. We will review trees and tree algorithms in the next chapter. Various books provide a much more thorough presentation and analysis of concepts reviewed in this chapter including Harary [4], Bondy and Murty [1, 2] and West [5].

Exercises

1. For the graph of Fig. 11.22,

   a. What is the degree of each vertex?
   b. What is the degree sequence of this graph?
   c. Find a spanning subgraph of this graph.
   d. Find the complement of this graph.

2. Given a graph G with a degree sequence $D = (d_1, d_2, ..., d_n)$, work out the degree sequence of the complement $\overline{G}$ of this graph.
3. Work out the adjacency list, adjacency matrix and the incidence matrix of the graph of Fig. 11.23.
4. Find the diameter, radius and the girth of the graph of Fig. 11.23.

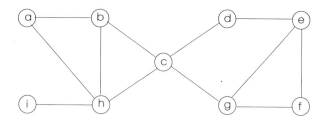

Fig. 11.22 Sample graph for Exercise 1

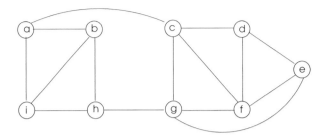

Fig. 11.23 Sample graph for Exercises 3 and 4

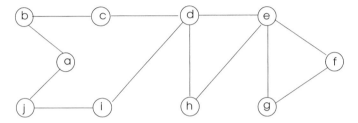

Fig. 11.24 Sample graph for Exercise 8

5. Find the diameter and radius of a complete bipartite graph $K_{m,n}$ in terms of m and n.
6. Show that for a graph with n vertices and m edges, $m \geq n$.
7. Show that the union of a simple undirected graph with n vertices and its complement is K_n.
8. Find an Eularian cycle if it exists in the graph of Fig. 11.24. Is there a Hamiltonian cycle in this graph? If not, add new edges to this graph to have such a cycle.
9. Draw the simple undirected unweighted graph represented by the following adjacency matrix.

$$
\begin{array}{c@{\quad}c}
 & \begin{array}{cccccccccc} a & b & c & d & e & f & g & h & i & j \end{array} \\
\begin{array}{c} a \\ b \\ c \\ d \\ e \\ f \\ g \\ h \\ i \\ j \end{array} &
\left(\begin{array}{cccccccccc}
0 & 1 & 0 & 0 & 0 & 0 & 0 & 0 & 0 & 0 \\
1 & 0 & 1 & 0 & 0 & 0 & 0 & 0 & 0 & 0 \\
0 & 1 & 0 & 0 & 0 & 0 & 0 & 0 & 0 & 0 \\
0 & 1 & 0 & 0 & 1 & 1 & 0 & 0 & 0 & 0 \\
0 & 0 & 0 & 1 & 0 & 1 & 0 & 0 & 0 & 0 \\
0 & 0 & 0 & 1 & 1 & 0 & 1 & 0 & 1 & 0 \\
0 & 0 & 0 & 0 & 0 & 1 & 0 & 1 & 1 & 0 \\
0 & 0 & 0 & 0 & 0 & 0 & 1 & 0 & 1 & 0 \\
0 & 1 & 0 & 0 & 1 & 0 & 1 & 1 & 0 & 1 \\
0 & 1 & 0 & 0 & 0 & 0 & 0 & 0 & 1 & 0
\end{array}\right)
\end{array}
$$

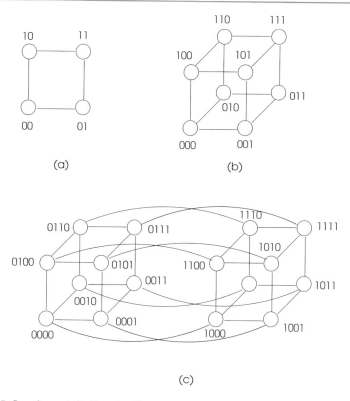

Fig. 11.25 Sample graph for Exercise 10

10. A *hypercube* H_n is a graph with n vertices such that each vertex is labeled with a binary numbers of length n for a total of 2^n vertices and the labels of adjacent vertices differ by 1 bit. Hypercubes of 2, 3 and 4 dimensions are depicted in Fig. 11.25a, b and c respectively with labels of vertices shown as binary numbers. Find the degree of each vertex, the number of edges and the diameter of a hypercube H_n in terms of the number of vertices n it contains.

References

1. Bollobas B (2002) Modern graph theory (Graduate texts in mathematics), Corrected ed. Springer, Berlin (2002). ISBN-10: 0387984887 ISBN-13: 978-0387984889
2. Bondy AB, Murty USR (2008) Graph theory (Graduate texts in mathematics), 1st Corrected ed. Springer, Berlin. Corr. 3rd printing 2008 edition (August 28, 2008), ISBN-10: 1846289696 ISBN-13: 978-1846289699
3. Diestel R (2010) Graph theory (Graduate texts in mathematics), 4th ed. Springer, Berlin. Corr. 3rd printing 2012 edition (October 31, 2010)

4. Harary F (1969) Addison Wesley
5. West D (2000) Introduction to graph theory. PHI learning, 2nd edn. Prentice Hall, Upper Saddle River
6. Gibbons A (1985) Algorithmic graph theory, 1st edn. Cambridge University Press, Cambridge. ISBN-10: 0521288819 ISBN-13: 978-0521288811
7. Skiena S (1990) Line graph. In: 4.1.5 in implementing discrete mathematics: combinatorics and graph theory with mathematica. Addison-Wesley, Reading, 128 and 135–139

Trees and Traversals

12

A tree is a graph with no cycles. Applications of trees are various; organization of an establishment, a family genealogical relationships can all be represented by a tree. Trees also find a number of applications in computer science, a fundamental usage is the representation of data. We start this chapter with the terminology and properties of trees. We then look at ways of traversing trees and describe specific tree types. In the second part of the chapter, our focus is on methods of tree construction from a general graph. Two basic methods for unweighted graphs are the breadth-first-search and the depth-first-search as we review. For weighted graphs, building an MST tree has many real-life applications as we will see.

12.1 Definitions and Properties

A *tree* is a connected graph without any cycles as noted and a *forest* is a graph with no cycles. A *rooted tree* has a designated vertex called the *root* and an *unrooted tree* has no root vertex. Each one of the following statements is adequate to define a tree.

- A tree is connected and has $n - 1$ vertices.
- Any two vertices of a tree are connected by a unique path.
- Each edge of a tree is a bridge, thus, removal of any edge of a tree disconnects the tree.

We have few definitions to be able to analyze the tree structure. A general rooted tree structure is depicted in Fig. 12.1. Let the root of such a tree be r and v be a vertex in somewhere in the middle of this tree. The first vertex u that is on the path to the root is named as the *parent* of vertex v. Note that there is only one such vertex

© Springer Nature Switzerland AG 2021
K. Erciyes, *Discrete Mathematics and Graph Theory*, Undergraduate Topics in Computer Science, https://doi.org/10.1007/978-3-030-61115-6_12

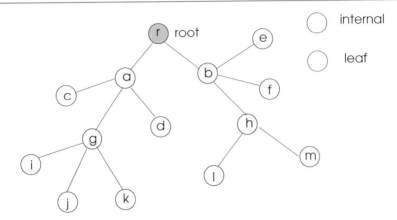

Fig. 12.1 A general tree structure

as otherwise we would have a cycle distorting tree property and v is called the *child* of vertex u. A vertex in a tree may have more than one child but can only have one parent as noted. A vertex without any children is called a *leaf* as shown by empty circles in the figure. Vertices with the same parent are called *siblings*. A vertex that is not a leaf and not a root is an internal vertex shown in gray in the figure. Such a vertex has one or more children and a parent as vertex g in the figure. The distance of a vertex to the root in a rooted tree is called its *level*. The level of the vertex g in the figure is 2 since it has a path of two edges to reach the root. The *depth* or the *height* of a tree is the largest of all levels in the tree which is 3 for this example.

We can have a simple recursive procedure that returns the level of a vertex in a tree as shown in Algorithm 12.1. The path to the root from vertex v is traversed recursively until root is encountered and each recursive call increments the value of the level.

Algorithm 12.1 *Finding depth of a vertex*

1: **procedure** FIND_DEPTH(v)
2:     **if** $v = root$ **then**
3:         **return** 0
4:     **end if**
5:     **return** $1 +$ FIND_DEPTH($parent(v)$)
6: **end procedure**

12.2 Traversal Algorithms

Visiting vertices of a given tree in some order is called *tree traversal*. Such a process has the property that each vertex should be visited exactly once. We have two basic algorithms for general trees as *preorder* and *postorder* traversals.

12.2.1   Preorder Traversal

In this traversal method, a vertex is first visited and then all of its children are visited. The general steps of the preorder traversal of a tree is shown in Algorithm 12.2. We start this procedure from the root typically and output each vertex label when visited. Visiting the vertices of the tree of Fig. 12.1 with this method starting from the root r results in $r, a, c, g, i, j, k, d, b, h, l, m, f, e$ as the output. This algorithm requires $O(n)$ steps as each vertex is visited exactly once.

Algorithm 12.2 *Preorder_Traversal*

1: **procedure** PREORDER(v)
2:     **visit** v
3:     **for all** child u of v **do**
4:         *Preorder*(u)
5:     **end for**
6: **end procedure**

12.2.2   Postorder Traversal

In this traversal method, the children of a vertex are visited first and the vertex is visited last as in Algorithm 12.3. Visiting the vertices of the tree of Fig. 12.1 results in $c, i, j, k, g, d, l, m, h, f, e, b, r$ as the output. Time complexity of this algorithm is also $O(n)$ as each vertex is visited exactly once.

Algorithm 12.3 *Preorder_Traversal*

1: **procedure** PREORDER(v)
2:     **for all** child u of v **do**
3:         *Preorder*(u)
4:     **end for**
5:     **visit** u
6: **end procedure**

12.3   Binary Trees

A *binary tree* is a tree where every vertex has at most two children. The child of a vertex on the left of a parent is termed *left child* and the one on the right is termed the *right child*. Every node of a *complete binary tree* has 0 or two children and each

Fig. 12.2 A binary tree

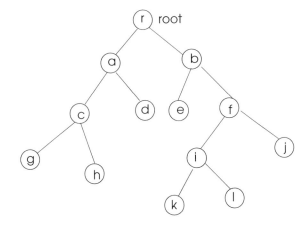

internal vertex and the root has exactly two children in such a tree. A complete binary tree is shown in Fig. 12.2.

We need to modify the preorder traversal for a binary tree taking left and right children into account as shown in Algorithm 12.4 The vertex is visited first, then its left subtree and then its right subtree in this method. Running of this algorithm in the binary tree of Fig. 12.2 starting from the root results visiting the vertices $r, a, c, g, h, d, b, e, f, i, k.l$ consecutively.

Algorithm 12.4 *Preorder binary tree traversal*

1: **procedure** PREORDER(v)
2:    *visit(v)*
3:    *Preorder($v.left\_subtree$)*
4:    *Preorder($v.right\_subtree$)*
5: **end procedure**

The postorder algorithm for a binary tree is modified similarly as in Algorithm 12.5 by fist visiting the left subtree of a vertex, then its right subtree and then the vertex. The vertices visited using this algorithm starting from the root in Fig. 12.2 are $g, h, c, d, a, e, k, l, i, f, b, r$.

Algorithm 12.5 *Postorder binary tree traversal*

1: **procedure** POSTORDER(v)
2:    *Postorder($v.left\_subtree$)*
3:    *Postorder($v.right\_subtree$)*
4:    *visit(v)*
5: **end procedure**

Inorder Traversal

Inorder traversal of a binary tree involves visiting the left subtree of a node first, then the node and then the right subtree of the node. Note that left and right notions are meaningful only in a binary tree, thus, inorder tree traversal is associated with binary trees. The recursive algorithm for inorder traversal is shown in Algorithm 12.6. The visited vertices of the tree in Fig. 12.2 in sequence are $g, c, h, d, a, r, e, b, k, i, l, f$. Time complexity of this algorithm is $O(n)$ using the same reasoning for other tree traversals.

Algorithm 12.6 *Inorder traversal*

1: **procedure** INORDER(v)
2:     *Inorder($v.left\_subtree$)*
3:     **visit** v
4:     *Inorder($v.right\_subtree$)*
5: **end procedure**

12.4   Binary Search Trees

A *binary search tree* (BST) is basically a binary tree where each node of this tree stores a *key* and the key value of a node is more than or equal to all of the key values in its left subtree and less than or equal to all key values of nodes in its right subtree. This type of binary tree is used for efficient data storage in computers. A BST is shown in Fig. 12.3.

A BST may be used to store data in non-contiguous locations in computer memory. Each node of the BST then has two pointers to the left node and to the right node where a pointer is basically a memory address. Each node of the BST typically will have a key, a value and two pointers for the left and right nodes as shown in Fig. 12.4. The leaves of the BST have *null* left and right pointers as they do not show any other node.

There are various operation that can be performed on a BST; we will review three such operations, searching a key value in a BST, finding the maximum value and the minimum key value stored in a BST. Searching a key value is shown as a recursive procedure in Algorithm 12.7. We assume each node of the BST is a tuple $a(left, value, right)$ where a is the label of the node, *left* is the left pointer, *value* is the value stored at a node and *right* is the right pointer. This procedure inputs the root node and a key value *val* and searches the tree until that value or a leaf node is encountered, thus, returning *null* pointer means the value is not found in the BST. We make use of the BST property such that if the value is less than the current node value, we go left; otherwise we go right in the subtrees.

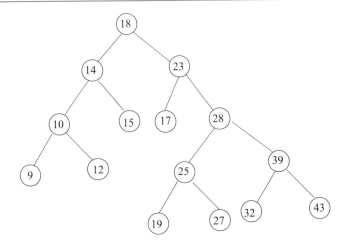

Fig. 12.3 A binary search tree

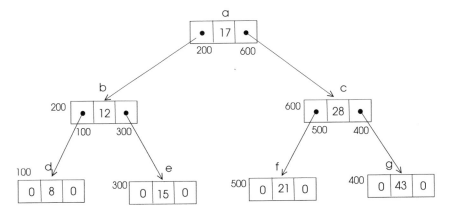

Fig. 12.4 A binary search tree in memory

Algorithm 12.7 *BST search*

1: **procedure** *BST_Search(node, value)*
2:     **if** *node = null* or *kvalue = node.val* **then**
3:         **return** *node*
4:     **end if**
5:     **if** *value < node.val* **then**
6:         **return** *BST_Search(node.left, value)*
7:     **else**
8:         **return** *BST_Search(node.right, value)*
9:     **end if**
10: **end procedure**

It may also be of interest to find the minimum value and the maximum value stored in a BST. The BST property means the minimum value stored in the tree is at farthest and lowest left node which can be accessed recursively by continuously going left in the tree until a *null* pointer is encountered as shown in Algorithm 12.8 and this is the node with the minimum value.

Algorithm 12.8 *BST minimum value*

1: **procedure** *BST_Min(node)*
2:     **while** *node.left* $\neq$ *null* **do**
3:         *node* $\leftarrow$ *node.left*
4:     **end while**
5:     **return** *node.value*
6: **end procedure**

The maximum value of a BST can be attained similarly to that of finding the minimum value but this time we go as right as possible until the node at the farthest right is reached as depicted in Algorithm 12.9.

Algorithm 12.9 *BST maximum value*

1: **procedure** *BST_Max(node)*
2:     **while** *node.right* $\neq$ *null* **do**
3:         *node* $\leftarrow$ *node.right*
4:     **end while**
5:     **return** *node.value*
6: **end procedure**

12.5   Depth-First-Search

A *depth-first-search* tree of a graph is a spanning tree of a graph discovered by staring from a vertex and going as deep as possible by traversing the edges in no specific order. Consider a floor of a building consisting of rooms with doors opening to other rooms. We are asked to visit every room in this floor starting from an arbitrary room and are given a ball of string and a piece of chalk. The floor is a graph, each room is a vertex and each door is an edge opening to another vertex (room) of the graph in our analogy. The algorithm to visit all of the rooms is to pull the string along as we move, whenever we enter a room through a door, mark the door with the chalk and whenever we come to a room with no doors or having all doors marked, we return to the room where we come from by rolling up the string.

A recursive algorithm to discover a DFS tree of a graph may be designed with the following principle. We start from a vertex u, mark it as visited and select one of the unmarked neighbors say v of u, set u as the parent of u and implement the same procedure, that is, mark v as visited, and select an unmarked neighbor of v say w and repeat. When there are no unvisited neighbors of a vertex, we return to the parent of the vertex. We also record the first time of visiting a vertex u and the last time when we need to return to the parent of u since it has no more unvisited neighbors. All vertices of a connected graph will be visited with this procedure, however, for a disconnected graph, we need to invoke the recursive procedure for each component.

The data structures to implement this algorithm are the boolean array $Visited[n]$ showing whether a vertex is visited or not, $Pred[n]$ showing the predecessor of a vertex in the BFS tree, an array $d[n]$ showing the first visit times of vertices and an array $f[n]$ showing the last visit times of vertices. The recursive DFS algorithm is shown in Algorithm 12.10. Note that call to this procedure at line 10 is invoked for each component of the graph as a single call which will result visiting all vertices of a component.

Algorithm 12.10 *DFS_Recursive_Forest*

1: **Input** : $G(V, E)$, directed or undirected graph
2: **Output** : $Pred[n]$; $d[n]$, $f[n]$ ▷ place of a vertex in DFS tree and its visit times
3: **int** *time* ← 0; **boolean** $Visited[1..n]$
4: **for all** $u \in V$ **do**   ▷ initialize
5:      $Visited[u] \leftarrow false$
6:      $Pred[u] \leftarrow \perp$
7: **end for**
8: **for all** $u \in V$ **do**
9:      **if** $Visited[u] = false$ **then**
10:         $DFS(u)$ ▷ call for each connected component
11:     **end if**
12: **end for**
13:
14: **procedure** $DFS(u)$
15:     $Marked[u] \leftarrow true$
16:     $time \leftarrow time + 1$; $d[u] \leftarrow time$ ▷ first visit
17:     **for all** $(u, v) \in E$ **do**   ▷ visit neighbors
18:         **if** $Visited[v] = false$ **then**
19:             $Pred[v] \leftarrow u$
20:             $DFS(v)$
21:         **end if**
22:     **end for**
23:     $time \leftarrow time + 1$
24:     $f[u] \leftarrow time$ ▷ return visit
25: **end procedure**

A DFS traversal of a sample forest is shown in Fig. 12.5 with tree edges shown in bold lines pointing to predecessors and the first and last visit times of a vertex

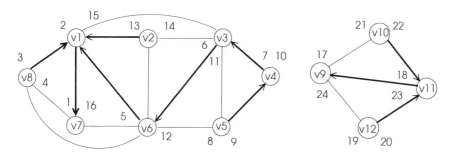

Fig. 12.5 DFS traversal of a sample graph with two components

is shown next to it with smaller number and larger number denoting the start and finish times respectively. We never traverse an edge between two marked edges, thus, the traversed edges form a tree. Since all vertices are visited, the tree formed is a spanning tree of the graph. The termination condition for the DFS procedure is returning to the first vertex it is called from and the algorithm finishes when all components are traversed, each with a single call to the DFS procedure which calls itself until all vertices in a component are visited. Note that selection of an unvisited vertex when done arbitrarily may result in different DFS tree structures. The time taken for this algorithm is $\Theta(n + m)$ in total; $\Theta(n)$ time for initialization and $\Theta(m)$ time for inspecting each edge twice from its endpoints. DFS algorithm is one of the building blocks of graph algorithms and is used extensively for more complex graph problems.

12.6 Breadth-First Search

The breadth-first search (BFS) algorithm is commonly used as the DFS algorithm to traverse the vertices of a graph. This time, *all* of the neighbors of an initial vertex u are visited, then all of the neighbors of these neighbors are visited and so on until all vertices of the graph are visited. The traversed edges at the end of the BFS algorithm is a spanning tree of the graph as in the DFS. This algorithm in an unweighted graph provides the distance of a vertex v in tree to the starting vertex u as the number of hops between u and v in the BFS tree. Whenever a vertex v is visited, it is assigned a value denoted by *level(v)* which is its distance to the starting vertex.

Considering the design of a BFS algorithm for a graph, we will input the graph and a starting vertex s to the procedure and we need the predecessor relationship in the array $Pred[n]$ and the level of a vertex in the BFS tree in the array $Levels[n]$ as the output as shown in Algorithm 12.11. We keep the neighbors of the current vertex under consideration in the queue Q and whenever a vertex from the queue is removed, its level is assigned only if this value is not assigned before, thereby preventing visiting the same vertex more than once.

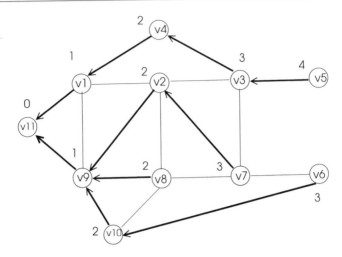

Fig. 12.6 BFS traversal of a sample connected graph

Algorithm 12.11 *BFS*

1: **Input** : $G(V, E), s$ ▷ undirected, connected graph G and a source vertex s
2: **Output** : $Level[n]$ and $Pred[n]$ ▷ levels and predecessors of vertices in BFS tree
3: **for all** $v \in V \setminus \{s\}$ **do**   ▷ initialize all vertices except source s
4:      $Level[v] \leftarrow \infty$
5:      $Pred[v] \leftarrow \perp$
6: **end for**
7: $Level[s] \leftarrow 0$ ▷ initialize source s
8: $Pred[s] \leftarrow s$
9: $Q \leftarrow s$
10: **while** $Q \neq \emptyset$ **do**   ▷ do until Q is empty
11:     $v \leftarrow deque(Q)$ ▷ deque the first element u
12:     **for all** $(u, v) \in E$ **do**   ▷ process all neighbors of u
13:         **if** $Level[u] = \infty$ **then**
14:             $Level[u] \leftarrow Level[v] + 1$
15:             $Pred[u] \leftarrow v$
16:             $enque(Q, u)$
17:         **end if**
18:     **end for**
19: **end while**

Running of this algorithm is displayed in Fig. 12.6 starting from vertex v_{11} with BFS tree edges shown in bold pointing to predecessors of a vertex and the level of a vertex is shown next to it. The order of enqueueing the neighbors of a vertex v in the queue may be done arbitrarily or based on some criteria as the node identifiers. The structure of the BFS depends on this order and we may obtain different BFS trees based on this selection. The initialization of the BFS algorithm takes $O(n)$ time and each edge is explored twice resulting in a total time of $O(n + m)$.

12.7 Spanning Trees

A *spanning tree* of a graph $G = (V, E)$ is a subgraph $T = (V, E')$ of G such that T is a tree and contains all vertices of G. We will review spanning tree construction of a graph as unweighted and weighted cases.

12.7.1 Unweighted Spanning Trees

Every connected graph has at least one spanning tree. The number of spanning trees of a graph G is denoted by $\tau(G)$. The number of spanning trees of the complete graph K_n which has distinct vertices is given by [2],

$$\tau(K_n) = n^{n-2} \tag{12.1}$$

For example, K_3 has 3 and K_4 has 16 such spanning trees when each vertex has a distinct label. A simple algorithm to construct a spanning tree of a graph is designed by initializing the spanning tree T to be formed by selecting an arbitrary edge (u, v) and including it in T and then always selecting edges with one endpoint in T and the other endpoint outside T, thereby preventing any cycle between selected edges. We maintain the acyclic property of a tree this way. This algorithm is depicted in Algorithm 12.12. Termination of the algorithm is when all vertices are included in T to obey spanning tree property.

Algorithm 12.12 *ST_Contsruct*

1: **Input** : $G = (V, E)$
2: **Output** : A spanning Tree T of G
3: $T \leftarrow$ an arbitrary edge (u, v)
4: $V' \leftarrow \{u, v\}$
5: **while** $V' \neq V$ **do**
6:     **select** any outgoing edge (u, v) from T with $u \in T \wedge v \notin T$
7:     $T \leftarrow T \cup \{(u, v)\}$
8:     $V' \leftarrow V' \cup \{v\}$
9: **end while**

The working of this algorithm is depicted in Fig. 12.7 where spanning tree edges are shown by bold lines and the order of inclusion of each edge included in spanning tree T is shown by a number. Thus, edges (b, g), (b, c), (h, g), (c, d), (h, a), (c, e) and (f, g) are included in sequence in T. This algorithm is correct since the structure we obtain in the end is acyclic, therefore a tree, and we ensure that all vertices are contained in the edges of the tree in line 5 of the algorithm which means the tree obtained is a spanning tree. Time complexity is simply $O(m)$ which is the number of edges of the graph.

Fig. 12.7 Construction of a
spanning tree of a small
sample graph using
Algorithm 12.12

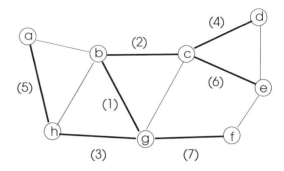

12.7.2   Minimum Spanning Trees

We can search a spanning tree in a weighted graph which may represent a communication network. In this case, our aim commonly is to build a spanning tree of the
graph which has a minimum total weight.

Definition 12.1 (*Minimum spanning tree*) A minimum spanning tree (MST) T of a
weighted graph $G = (V, E, w)$ with $w : E \rightarrow \mathbb{R}$ is the minimum weight spanning
tree of G among all spanning trees of G. Formally,

$$w(T) = \sum_{(u,v) \in T} w(u, v) \tag{12.2}$$

Let $G = (V, E, w)$ be a connected, weighted, undirected graph. The following
properties of G proofs of which can be found in [1] form the basis of some MST
algorithms.

- *Uniqueness*: G has a unique MST if the edge weights of G are distinct.
- *Cut Property*: Let $H \in E$ be a subset of some MST of G and let $(S, V \setminus S)$ be any
  cut in G that does not contain any edge in H. Then, the least weight edge (u, v)
  that crosses this cut is contained in some MST of G.
- *Cycle Property*: Let C be any cycle in G and (u, v) be the maximum weight edge
  in C. Then, (u, v) is not contained in any MST of G.

12.7.2.1   Prim's Algorithm

Prim's algorithm to find the MST of a weighted graph $G = (V, E, w)$ is based on
the cut property as follows. Let T' be a subset of an MST of G; then, a minimum
weight outgoing edge (MWOE) from any vertex of T' is contained in the final MST
of G. The MWOE has the least weight edge among all edges that have one endpoint
in T' and the other endpoint outside T'. A partial MST T' is shown in bold lines in
Fig. 12.8 and the MWOE is in dashed line. Note that edge (v_8, v_9) is not an MWOE
of T' since both of its endpoints are contained in T'.

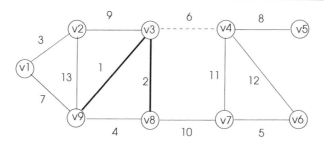

Fig. 12.8 A partial MST and MWOE

Prim's algorithm selects and arbitrary vertex r as the root of the MST to be formed, includes this vertex in MST T and thereafter, repeatedly selects the MWOE from the current MST fragment, includes MWOE in the fragment and continues until all vertices of G are included in the MST as shown in Algorithm 12.13.

Algorithm 12.13 *Prim's MST algorithm*

1: **Input** : $G = (V, E, w)$ ▷ a weighted graph
2: **Output** : $T = (V, E_T)$ ▷ MST of G
3: $V' \leftarrow \{r\}$
4: $T \leftarrow \emptyset$
5: **while** $V' \neq V$ **do**   ▷ continue until all vertices are visited
6:     **select** MWOE (u, v) of T with $u \in T$ and $v \in G \setminus T$
7:         $V' \leftarrow V' \cup \{v\}$
8:         $E_T \leftarrow E_T \cup \{(u, v)\}$
9: **end while**

The working of Prim's algorithm in an example graph is shown in Fig. 12.9. The main operation in this algorithm is the selection of MWOE. Time complexity of Prim's algorithm is $O(m \log n)$ using suitable data structures [1].

12.7.2.2   Kruskal's Algorithm

Kruskal's algorithm works differently by first sorting the edges of a graph $G = (V, E, w)$ with respect to their weights in non-decreasing order and the MST T contains no edges initially. Then, starting from the lightest edge, an edge in the list is included in T as long as this edge does not make a cycle with edges already contained in T. The algorithm to perform this procedure is shown in Algorithm 12.14.

The running of this algorithm for the same graph of Fig. 12.9 is shown in Fig. 12.10 The sorting process takes $O(m \log m)$. Note that the edge (a, f) is not included in T as it would form a cycle with the existing T edges if included which is the dominant time for this algorithm.

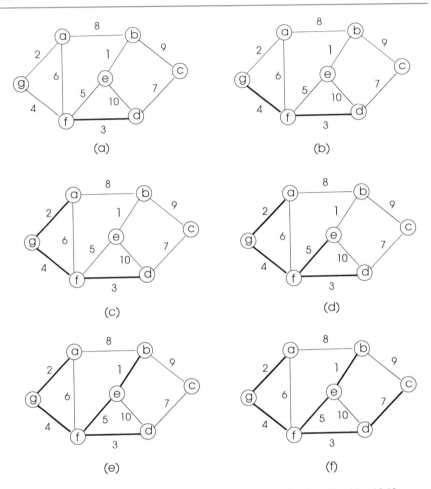

Fig. 12.9 Construction of a spanning tree of a small sample graph using Algorithm 12.13

Algorithm 12.14 *Kruskal_MST*

1: **Input** : $G = (V, E, w)$ ▷ a weighted graph
2: **Output** : $T = (V, E_T)$ ▷ MST of G
3: $T \leftarrow \emptyset$
4: $Q \leftarrow$ sorted edges in nondecreasing weights of E
5: **while** $Q \neq \emptyset$ **do**   ▷ check all edges
6:     **remove** the first edge (u, v) from Q
7:     **if** (u, v) does not make a cycle with current edges in T **then**
8:         $T \leftarrow T \cup (u, v)$
9:     **end if**
10: **end while**

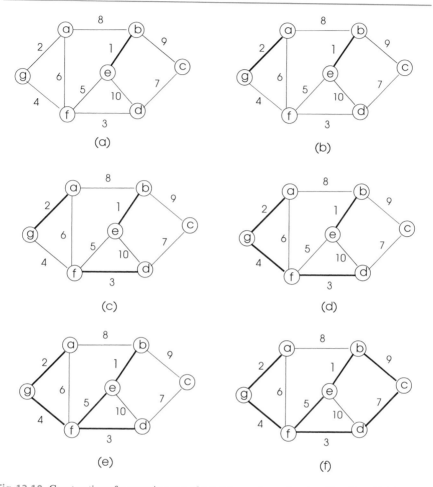

Fig. 12.10 Construction of a spanning tree of a small sample graph using Algorithm 12.14

12.7.2.3   Reverse Deletion Algorithm

This algorithm works by first sorting the edges of a graph $G = (V, E, w)$ with respect to their weights in non-increasing order and the MST T contains all edges of G initially. Then, starting from the heaviest edge, an edge in the T is deleted from T as long as this edge does not disconnect the current T since T must be connected. The main idea of this algorithm is the fact that a heaviest weight edge in a cycle can not be part of any MST of a weighted graph as defined by the cycle property. The algorithm to perform this procedure is shown in Algorithm 12.15.

We apply this algorithm in the graph as we did for Prim's and Kruskal's algorithms and the final MST obtained is the same as shown in Fig. 12.11. Note that we do not delete edge (c, d) since doing so leaves the current T unconnected. Also, deletion of any edge with weight less than 6 leaves G disconnected and thus, these edges are not removed. The running time is dominated by the sorting procedure which takes $O(m \log m)$ steps.

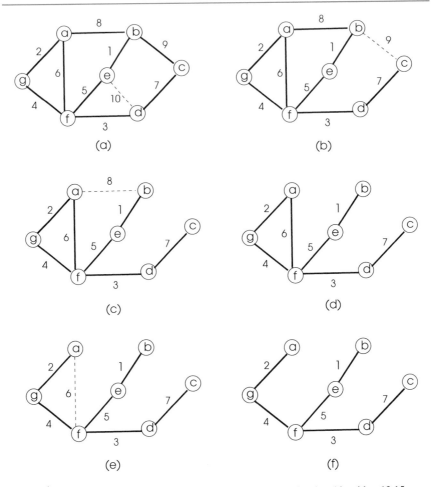

Fig. 12.11 Construction of a spanning tree of a small sample graph using Algorithm 12.15

Algorithm 12.15 *Reverse_Delete_MST*

1: **Input** : $G = (V, E, w)$
2: **Output** : MST T of G
3: $T \leftarrow e$
4: $Q \leftarrow$ sorted edges in noninreasing weights of E
5: **while** $Q \neq \emptyset$ **do**   ▷ check all edges
6:    **remove** the first edge (u, v) from Q
7:    **if** deleting (u, v) from T does not leave T disconnected **then**
8:       $T \leftarrow T - (u, v)$
9:    **end if**
10: **end while**

12.8   Review Questions

1. Compare briefly a preorder and postorder traversal of a general tree.
2. What is a binary tree?
3. What is the difference between preorder and postorder traversals of a general tree and a binary tree?
4. Why inorder traversal is meaningful only in a binary tree?
5. What is a binary search tree and where can it be used?
6. Describe a method to build a spanning tree of an unweighted graph.
7. Define a minimum spanning tree of a weighted graph.
8. What makes Prim's MST algorithm correct?
9. Compare Kruskal's MST algorithm with the Reverse Delete MST algorithm.
10. Why is Kruskal's algorithm correct?
11. What makes Reverse Delete MST algorithm work correctly?

12.9   Chapter Notes

We first reviewed basic tree properties and traversals of a tree in this chapter. Preorder and postorder traversals are applicable to all trees. A binary tree has nodes having at most two children. The inorder traversal procedure may be defined for a binary tree since a binary tree has the notion of left and subtrees. A binary search tree has elements ordered such that any element in a subtree of a node v in such a tree has a value less than or equal to all values in v and the right subtree of v.

We then reviewed two basic graph traversal algorithms; DFS and BFS. Vertices in a DFS algorithm are visited by going as deep as possible and the BFS algorithm visits vertices in layers thereby finding shortest distances to the starting vertex in an unweighted graph. Both of these algorithms produce spanning trees and may be used as building blocks of more complex graph problems such as finding connectivity and clustering which is a method of grouping closely related vertices of graph.

Lastly, we looked at constructing a general spanning tree of an unweighted undirected graph and minimum spanning tree (MST) algorithms. We analyzed Prim's, Kruskal's and Reverse Delete MST algorithms which have similar time complexities. The MST problem is one of the most investigated problems in Computer Science as it has numerous applications in various disciplines.

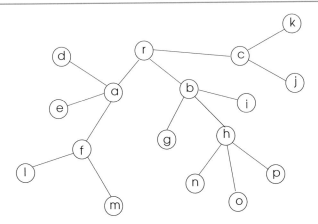

Fig. 12.12 Sample tree for Exercise 2

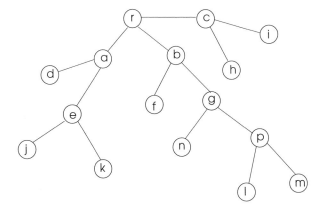

Fig. 12.13 Sample tree for Exercise 4

Exercises

1. Prove that every graph G with n vertices and $n - 1$ edges is a tree using the contradiction method.
2. Work out the order of the visited vertices using preorder and postorder traversals of the general tree in Fig. 12.12.
3. Prove that every tree is a bipartite graph.
4. Find the list of visited vertices of the binary tree in Fig. 12.13 using inorder traversal.
5. Propose an algorithm to insert a value x in a binary search tree
6. Find the MST of the weighted graph of Fig. 12.14 by Prim's algorithm.
7. Work out a possible DFS tree in the graph of Fig. 12.15 starting from vertex v_1. Show the start and finish times for each vertex and the predecessor of each vertex in the tree formed.

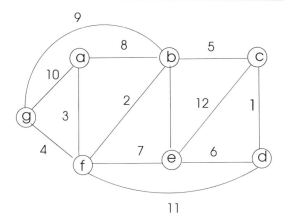

Fig. 12.14 Sample graph for Exercises 6, 9 and 10

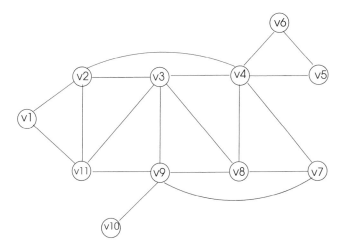

Fig. 12.15 Sample graph for Exercises 7 and 8

8. Find a BFS of the graph in Fig. 12.15 starting from vertex v_9 and show the level and predecessor of each vertex in the tree formed.
9. Find the MST of the weighted graph of Fig. 12.14 by Kruskal's algorithm.
10. Find the MST of the weighted graph of Fig. 12.14 by the Reverse-Delete algorithm.

References

1. Erciyes K (2018) Guide to graph algorithms: sequential, parallel and distributed. Springer texts in computer science series
2. Cayley A (1857) On the theory of analytical forms called trees. Philos Mag 4(13):172–176

Subgraphs

13

In this chapter, we will review few special subgraphs that find numerous real-life applications. We start with the clique which is a fully connected graph. Finding cliques in a graph is needed to discover closely related nodes of the graph which may be proteins, network nodes or even persons. We then look at a fundamental problem called matching in a graph which is a set of disjoint edges. Independent sets, dominating sets and vertex cover each provide a subset of vertices of a graph with a specific property. We conclude this chapter with another well-known graph problem: vertex coloring. where each vertex should be colored with a different color than its neighbors. All of these problems except matching are NP-hard problems defying solutions in polynomial time.

13.1 Cliques

A *clique* is a fully connected graph, that is, there is an edge between any pair of vertices in a clique. Finding cliques in graphs have various implications, for example, finding closely related friends in a social network when the social network is represented by a graph. Cliques of size 2, 3 and 4 are depicted in Fig. 13.1.

A graph itself may not be a clique but may have a subgraph that is a clique. A clique of a graph is defined as follows.

Definition 13.1 (*Clique*) A *clique* of a graph $G = (V, E)$ is a subset V' of its vertices with an edge between any pair of vertices in V'. In other words, a clique is an induced complete subgraph of G. Formally, given $G = (V, E)$ and $V' \subseteq V$; $(u, v) \in E \ \forall u, v \in V'$.

A maximal clique C of a graph G has the highest order among all cliques of G. The order of a maximum clique of G is denoted by $\omega(G)$ and finding this parameter

© Springer Nature Switzerland AG 2021
K. Erciyes, *Discrete Mathematics and Graph Theory*, Undergraduate Topics in Computer Science, https://doi.org/10.1007/978-3-030-61115-6_13

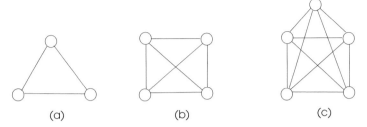

(a)   (b)   (c)

Fig. 13.1 Cliques of order **a** 3, **b** 4 and **c** 5

Fig. 13.2 Cliques of a
sample graph

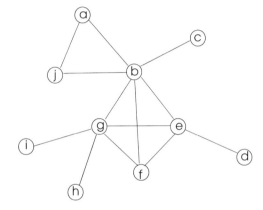

is an NP-hard problem. The sample graph of Fig. 13.2 has 5 cliques of order 3 and
a maximum clique of order 4 with vertices (b, e, f, g) in the middle as subgraphs.

13.2   Matching

A *matching* of a graph G is a subset of edges of G that do not have any common
endpoints. Matching finds various applications such as in bioinformatics to find
similarities between biological structures and in telecommunications when allocating
channel frequencies.

13.2.1   Unweighted Matching

Definition 13.2 (*Unweighted matching*) A matching of an unweighted graph $G =
(V, E)$ consists of a set of edges $E' \subseteq E$ such that any edge in E' does not share
endpoints with any other edge in E'. A maximal matching (MM) of a graph G is a
matching of G which can not be enlarged any further. A maximum matching (MaxM)
of G has the largest size among all matchings of G.

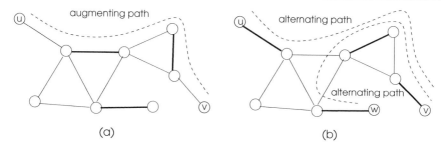

Fig. 13.3 **a** A MM of a graph, **b** a MaxM of the same graph shown by bold edges

An MM and MaxM of a sample graph of sizes 2 and 3 respectively are depicted in Fig. 13.3. Finding MaxM of a graph is one of the rare graph problems that has solutions in polynomial time. We can have a simple algorithm to find maximal matching in a graph $G = (V, E)$, consisting of the following steps.

1. Maximal matching $M \leftarrow \emptyset$
2. **while** $E \neq \emptyset$
3.   Select an edge $e \in E$ at random.
4. $M \leftarrow M \cup e$.
5.   Remove e and all its adjacent edges from E.

Correctness of the algorithm is evident since we obey matching property by removing all adjacent edges of the selected edge and the time complexity of this algorithm is $O(m)$ since we iteratively remove all unmatched edges. Running of this algorithm is depicted in Fig. 13.4 using the same graph of Fig. 13.3. Selected edges at each iteration are shown in bold and removed edges are shown by dashed lines, and the final graph with all matching edges is shown in (d). The matching obtained is maximum with a size of 4.

13.2.1.1 Augmenting Path

We need to define some concepts before forming a more efficient algorithm to find matching in an unweighted graph than the greedy method described. Our aim is to find a maximum matching of a graph in polynomial time. A *matched edge* of a matching M in a graph G is an edge contained in M and an unmatched edge of G is called an *unmatched* edge. An *M-alternating path* of M is a path that alternates between matched edges in M and unmatched edges. Such two M-alternating paths of the maximum matching in Fig. 13.3b is shown in dashed lines between vertices u and v, and vertices w and v.

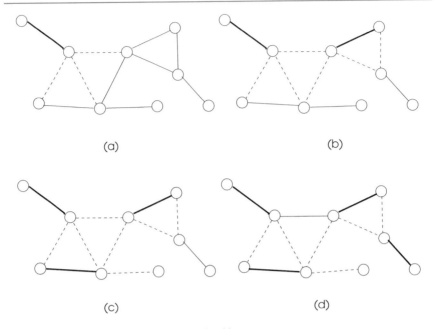

(a)   (b)

(c)   (d)

Fig. 13.4 Iterations of the greedy matching algorithm

Definition 13.3 (*Augmenting path*) An *M-augmenting path* of a matching M in a graph G is an alternating path that starts and ends in an unmatched edge in G.

An augmenting path is depicted Fig. 13.3a in dashed lines between vertices u and v. An M-augmenting path in a graph G that has k edges has $k + 1$ unmatched edges and thus has $2k + 1$ edges in total. In the example augmenting path, $k = 2$ and there are 5 edges in this augmenting path.

Remark 22 Let M be a matching of a graph G and P an M-augmenting path in G. Then $M \oplus P$ is again a matching in G with a cardinality one more than that of M.

Note that $M \oplus P$ is the symmetric difference of sets M and P which contains elements that belong to only one of these sets. In practice, this operation means switching the matched edges to unmatched and unmatched edges to matched in P as shown in the path in Fig. 13.3b. The following theorem due to Berge forms the basis of various matching algorithms [1].

Theorem 23 *Given a graph with matching M, M is maximum if M does not contain any augmenting paths.*

If there was an augmenting path P, we would increase the size of the matching by taking the symmetric difference of M and P. Based on this theorem, we can find a maximum matching in a graph as shown by the simple algorithm below, however, finding augmenting paths of a matching is not trivial as we will see.

1. Let M be some initial matching in G obtained by the greedy or another method.
2. **while** $\exists$ an augmenting path P in M
3. $\quad$ $M \leftarrow M \oplus P$.

13.2.2 Weighted Matching

Maximum weighted matching (MaxWM) for a weighted graph $G = (V, E, w)$ is defined as the matching with the total weight of matched edges among all matchings of G. Note that the size of a MaxWM of G may be smaller than some other matching that have less total weight. A greedy approach to find maximal weighted matching (MWM) may be formed as in the unweighted case, this time selecting always the largest weight edge as shown below. Sorting edges requires $O(m \log m)$ time and checking for edges to be removed from the queue takes $O(m^2)$ time as we have to do this for every edge selected. Thus, this algorithm has $O(m^2)$ time complexity in its naive form.

1. Maximal weighted matching $M \leftarrow \emptyset$
2. Sort the edges of G in descending order and store them in Q.
3. **while** $Q \neq \emptyset$
4. $\quad$ Remove the first element e from the queue.
5. $\quad$ $M \leftarrow M \cup e$.
6. $\quad$ Remove all adjacent edges of e from Q.

Running of this algorithm in a sample graph is shown in Fig. 13.5 with the selected edges shown in bold lines and removed edges shown by dashes lines. The final weighted matching is shown in (d) with a total weight of 30 which in fact is the maximum weighted matching of this graph.

Preis' Algorithm

The algorithm due to Preis is a greedy algorithm for weighted matching in weighted graphs [6]. A locally heaviest edge in this algorithm is defined as the edge with the largest weight among all of its adjacent edges. These edges are selected randomly at each iteration of this procedure, included in the matching and all of the adjacent edges are removed from the graph. A running of this algorithm is shown in Fig. 13.6 for the same graph of Fig. 13.5 with the final matching shown in (d) having a total weight of 28. Time complexity of this algorithm is given as $O(m \log n)$ with an approximation ratio of 2 [6].

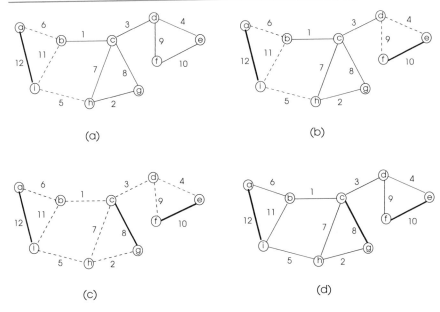

Fig. 13.5 Iterations of the greedy matching algorithm

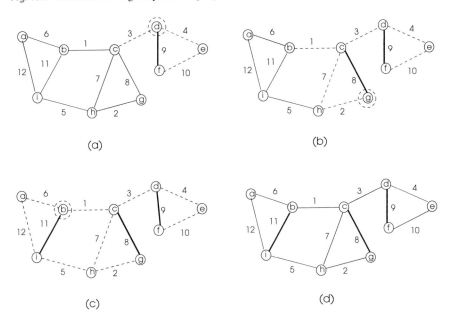

Fig. 13.6 Iterations of Preis' algorithm

13.2.3 Bipartite Graph Matching

A simple greedy algorithm to find a maximal weighted matching in a bipartite weighted graph can be designed to consist of the following steps.

1. **Input**: $G = (A \cup B, E)$
2. **Output**: Maximal weighted matching M of G
3. **sort** the edges of G in non-increasing order and store in a queue Q.
4. **while** $Q \neq \emptyset$
5.   **dequeue** edge (u, v) from Q.
6.   **add** (u, v) to M.
7.   **remove** all edges incident to u or v from Q
8. **end while**

This algorithm works correctly since we obey matching property by removing all adjacent edges to the matched edges from consideration. This algorithm first sorts the edges in non-decreasing weights and then selects a legal edge from the list in sequence. This algorithm has $O(m^2)$ time complexity in its naive form as the general weighted graph algorithm.

13.3 Independent Sets

Informally, an independent set (IS) of a graph is a subset of its vertices such that there are no edges between any pair of vertices in this set; a formal definition follows.

Definition 13.4 (*Independent set*) An independent set of a graph $G = (V, E)$ is the vertex set $V' \subseteq V$ such that for any $u, v \in V'$, $(u, v) \notin E$. A *maximal independent set* (MIS) of a graph G is an independent set of G which can not be enlarged any further. In other words, we can not add another vertex to MIS of a graph because this new vertex will be a neighbor vertex to one of the vertices in the MIS. The maximum independent set (MaxIS) of G is the independent set of maximum order among all independent sets of G.

The example graph in Fig. 13.7a shows an MIS of this graph in bold nodes. The graph in Fig. 13.7b however is an MaxIS of the same graph shown with bold nodes. Note that we can have a number of MISs or MaxISs of the same graph.

13.3.1 Algorithm

We can form a simple greedy algorithm to find the MIS of a graph G by iteratively selecting a vertex v to be included in the MIS and then removing v and all of its

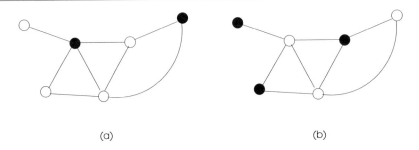

Fig. 13.7   **a** An MIS of a graph, **b** the MaxIS of the same graph

neighbors from this set to be searched in the next iteration as shown in Algorithm 13.1. Removing the neighbors of vertex v along with all incident edges on these neighbors ensures the IS property is maintained. The algorithm terminates when there are no more vertices left to be included in the IS, thus, this algorithm works correctly to find the MIS of a graph. The total time taken is $O(n)$ as the number of possible iterations of the *while* loop. Finding a MaxIS of a graph however is an NP-hard problem and there are no known algorithms to find this set in polynomial time.

Algorithm 13.1 *Find_MIS*

1: **Input**: $G = (V, E)$:   ▷ undirected graph
2: **Output**: MIS V' of G
3: $S \leftarrow V$
4: $V' \leftarrow \emptyset$
5: **while** $S \neq \emptyset$ **do**
6:     **select** an arbitrary vertex $v \in S$
7:     $V' \leftarrow V' \cup \{v\}$
8:     $S \leftarrow \{S - N(v) \cup \{v\}\}$
9: **end while**

The execution steps of this algorithm on the sample graph of Fig. 13.7 is depicted in Fig. 13.8 where nodes selected for MIS in sequence are shown in bold and the deleted vertices along with their incident edges are shown by dashed lines. The obtained MIS is in fact another MaxMIS for this graph.

Let I be any IS of a graph $G = (V, E)$, thus, there is not an edge between any pair of vertices in I. Then, the complement of G, $\overline{G}$, will have all possible edges between the vertices of I, therefore forming a clique in $\overline{G}$. This relation is depicted in Fig. 13.9.

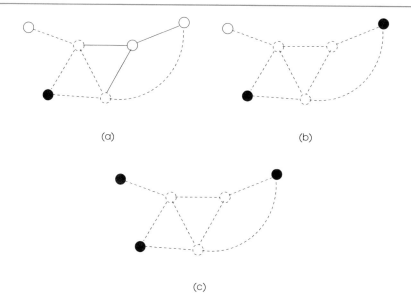

Fig. 13.8 Execution steps of Algorithm 13.1

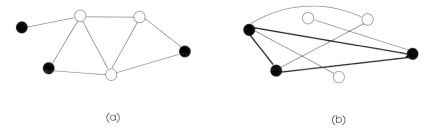

Fig. 13.9 **a** A sample graph with an independent set I shown in bold, **b** I is a clique in $\overline{G}$ same graph

13.4 Dominating Sets

A dominating set of a graph G is a subset of its vertices with the condition that every vertex in G is either in this set or a neighbor of a vertex in this set.

Definition 13.5 (*Dominating set*) A dominating set (DS) of a graph $G = (V, E)$ is the vertex set $V' \subseteq V$ such that for any $v \in V$, either $v \in V'$ or $v \in N(u)$ where $u \in V'$. A minimal dominating set (MDS) can not be reduced any further and a minimum dominating set (MinDS) of a graph G is a dominating set of G that has the least number of vertices among all dominating sets of G. In other words, a MDS of a graph G does not contain any dominating set as a proper subset.

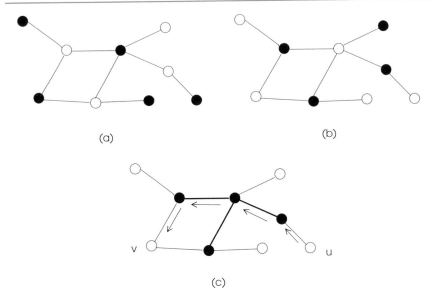

(a)

(b)

(c)

Fig. 13.10 a A MDS of a graph, **b** A MinDS of the same graph, **c** A connected MinDS of the same graph

An alternative definition of a dominating set can be stated as follows. Let $V' \subseteq V$ and $\forall v \in V - V', d(u, v) \le 1$ for any vertex $u \in V'$. Then V' is a dominating set of G.

Definition 13.6 (*Domination number*) The domination number $\gamma(G)$ of a graph G is the minimum cardinality of any dominating set of G.

There are two important concepts to be noted about dominating sets. The first one is we search for a minimal dominating set of a graph since our aim is to use a minimal number of nodes to be able to dominate every node of the graph. Many applications require this facility, for example, we may use a dominating set to form a backbone in a communication network such as a wireless sensor network. Secondly, unlike an independent set, the nodes of a dominating set may be adjacent. Such a dominating set with a path between any pair of nodes of the dominating set is called a connected dominating set (CDS). This is indeed the case when we need to find the dominating set in a communication network. A minimal dominating set in which taking out a node from this set results in disturbance of the dominating set property is depicted in Fig. 13.10a. A MinDS of the same graph is shown in (b) and lastly, a CDS of the same graph is shown in (c) which can be used as a backbone for communication. Edges shown between the nodes represent communication links and the node u sends a message to node v over the backbone formed by the CDS nodes. The algorithm performed by each CDS node is then check the destination node identifier v in the message, if node v is one of its dominated nodes, send the message to v and stop transmission. Otherwise, it simply broadcasts the message to all of its CDS neighbors. This method is effective since we do not need to broadcast the message to all nodes of the network.

13.4.1 Algorithm

Finding MinDS is an NP-hard problem with no known polynomial algorithms. However, we can have an algorithm to find MDS with the following considerations. Our coloring scheme denotes nodes in DS by *black*, nodes that are dominated by *grey* and any node that is not in DS or not dominated is *white*. Clearly, the algorithm designed should continue until no more white nodes left. Let the *span* of a node v to be the number of white neighbors v has including itself. Our basic heuristic is to always select the node with the highest number of span in the graph. This heuristic is sensible since our aim is to have a MDS, thus, we attempt to cover as many white nodes as possible at each step. Whenever a white node is selected to be in the MDS, the span values of its neighbors should be decremented. Algorithm 13.2 shows the operation of this algorithm with the described heuristic.

Algorithm 13.2 *Find_MDS*

1: **Input** : $G = (V, E)$ ▷ connected, unweighted graph
2: **Output** : MDS V' of G
3: $Color[n] \leftarrow white$
4: $V' \leftarrow \emptyset$
5: **for all** $u \in V$ **do**
6:     $Spans[u] \leftarrow |N(u)| + 1$
7: **end for**
8: **while** there is a node u with $Color[u] = white$ do **do**
9:     $v \leftarrow max(Spans)$
10:     $prev \leftarrow Color[v]$
11:     $Color[v] \leftarrow black$
12:     $V' \leftarrow V' \cup \{v\}$
13:     **for all** $w \in N(v)$ **do**
14:         **if** $prev = white$ and $Color[w] \neq black$ **then**
15:             $Spans[w] \leftarrow Spans[w] - 1$
16:         **end if**
17:         **if** $Color[w] = white$ **then**
18:             $Spans[w] \leftarrow Spans[w] - 1$
19:             $Color[w] \leftarrow grey$
20:             **for all** $k \in N(w)$ **do**
21:                 **if** $Color[k] \neq black$ **then**
22:                     $Spans[k] \leftarrow Spans[k] - 1$
23:                 **end if**
24:             **end for**
25:         **end if**
26:     **end for**
27:     $Spans[v] \leftarrow 0$
28: **end while**

Another important point about this algorithm is worth mentioning. We may color a *grey* node black if it has the highest span among all *white* or *grey* nodes. However, we do not consider *black* nodes as they are already in the MDS, thus span of a selected node is made zero in line 19 of the algorithm. Implementation of this algorithm in

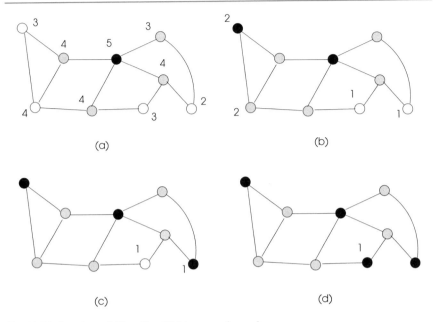

Fig. 13.11 Iterations of Algorithm 13.2 in a sample graph

a sample graph is depicted in Fig.13.11 with the initial span values at each iteration shown next to nodes and node identifiers may be used to break symmetries when there is a tie. The time complexity of this algorithm is $O(n)$ as in the case of a linear network and its approximation ratio is $\ln \Delta$ where Δ is the highest node degree of the graph under consideration [7]. A CDS may be obtained from a MDS by connecting the intermediate nodes between the nodes of the MDS using a suitable algorithm.

13.5 Coloring

Coloring refers to nodes or edges of a graph where colors are commonly represented by positive integers. Vertex coloring of a graph results in a color of a node that is different than the colors of its neighbors. Similarly, edge coloring is the coloring of edges such that no adjacent edges receive the same color.

13.5.1 Vertex Coloring

Definition 13.7 (*Vertex coloring*) Let C be the set of integers $1, ..., k$. A *vertex coloring* of a graph $G = (V, E)$ is a function $\phi : V \to C$ such that $\phi(u) \neq \phi(v)$ if $(u, v) \in E$.

Clearly we can color a graph G with n colors where n is the number of vertices of G. However, the aim of any coloring algorithm is to use as less colors as possible.

Definition 13.8 (*Chromatic number*) The chromatic number of a graph G denoted by $\chi(G)$ is the minimum number of colors to color graph G. A graph that has a chromatic number k is called a *k-chromatic graph* and a graph that has $\chi(G) \le k$ is said to be *k-colorable*.

Finding chromatic number of a graph is NP-hard with no known polynomial algorithms [5] and the value of this parameter is $O(\Delta)$ by Brooke's theorem [2], if graph is not complete or an odd cycle. A bipartite graph is 2-colorable since all vertices in each vertex set can have the same color. We can have various greedy algorithms using some heuristic to color the vertices of a graph. Let us form a template of an algorithm that can be used for various heuristics as shown in Algorithm 13.3 [3]. This algorithm selects a vertex v based on some property and colors it with the minimum legal color that does not conflict with the already assigned colors of the neighbors of v. Since we obey the vertex coloring property at each iteration, the algorithm works correctly.

Algorithm 13.3 *Coloring_Template*

1: **Input** : $G = (V, E)$
2: **Output** : $\phi : V \to C$ where $C = \{1, 2, ..., n\}$
3: **while** $V \ne \emptyset$ **do**
4:     **select** a vertex $v \in V$ according to some heuristic
5:     $\phi(v) \leftarrow$ the smallest legal color from C
6:     $V \leftarrow V \setminus \{v\}$
7: **end while**

We will describe two heuristics that can be used by this template; a label-based and a degree based heuristic. In the first case, each vertex is labeled with integers and the highest numbered vertex is selected at each iteration and colored with the minimum legal color. Implementation of this heuristic is depicted in Fig. 13.12 which shows the coloring of a sample graph with 4 colors.

The degree-based heuristic always selects the highest-degree vertex that is not assigned any color and colors it with the minimum legal color. The running of this algorithm in the same graph of Fig. 13.12 is shown in Fig. 13.13. The larger identifier is selected when uncolored vertices have the same degree. In this implementation, we can see that coloring of the same graph is achieved using 3 colors only.

13.5.2 Edge Coloring

The edge coloring of a graph G is the process of assigning different colors to the edges of G such that no two adjacent edge receives the same color. The colors are selected from integers $1, ..., k$ as in the vertex coloring problem.

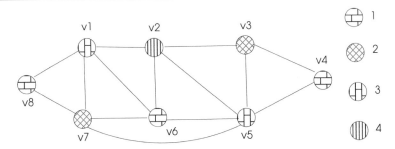

Fig. 13.12 Vertex coloring using vertex identifiers in a sample graph

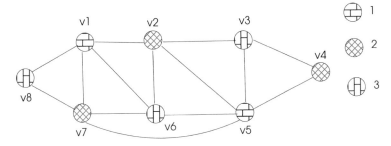

Fig. 13.13 Vertex coloring using vertex degrees of the sample graph of Fig. 13.13

Definition 13.9 (*Edge coloring*) Let C be the set of integers $1, ..., k$. An *edge coloring* of a graph $G = (V, E)$ is a function $\phi : E \rightarrow C$ such that for any two adjacent edges e_i and e_j, $\phi(e_i) \neq \phi(e_j)$.

A graph G is called *k-edge colorable* if there exists an edge coloring $\phi(G) : E \rightarrow C$ such that $|C| = k$. The edge chromatic number $\chi'(G)$ of a graph G is the minimum k value such that G is k-edge colorable. The *edge coloring problem* is defined as finding the minimum number of colors to edge-color a graph and this problem is also NP-hard as the vertex coloring problem.

Remark 13.1 The edge chromatic number of a graph G, $\chi'(G)$, is greater than or equal to the maximum degree $\Delta(G)$ of G. This observation follows from the fact that the edges incident to the maximum degree vertex of G need to be colored at least by Δ colors.

Remark 13.2 The chromatic number of a bipartite graph G, $\chi'(G)$ equals the maximum degree $\Delta(G)$ of G.

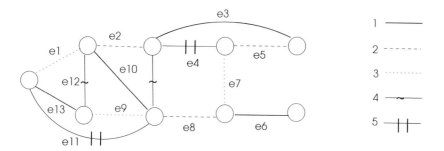

Fig. 13.14 Edge coloring of a sample graph using Algorithm 13.4

13.5.2.1 A Greedy Algorithm

A simple greedy algorithm which selects an uncolored edge at random and colors it with the minimum legal color that does not conflict with the already assigned colors of adjacent edges can be formed as shown in Algorithm 13.4

Algorithm 13.4 *Greedy_Ecolor*

1: **Input** : $G = (V, E)$
2: **Output** : $\phi' : E \rightarrow C$ where $C = \{1, 2, ..., 2\Delta - 1\}$
3: $E' \leftarrow E$
4: **while** $E' \neq \emptyset$ **do**
5:     **select** an edge $e \in E'$
6:     $\phi'(e) \leftarrow$ the smallest legal color from C
7:     $E' \leftarrow E' \setminus \{e\}$
8: **end while**

Running of this algorithm is depicted in Fig. 13.14. The edges are selected in order $e_{10}, e_2, e_6, e_8, e_4, e_{13}, e_3, e_7, e_1, e_5, e_{12}, e_7, e_{14}$ to result in 5-edge-coloring of this example graph. Time complexity of this algorithm is $\Theta(m)$ since we need to color an edge at each step.

13.5.2.2 Edge-Coloring from Matching

A matching of a graph G consists of nonadjacent edges, thus, edges of a distinct matching of G can be colored with the same color. The edge coloring of G is then a union of *disjoint* matchings. We can therefore iteratively select disjoint matchings and color edges in each matching with the same color as shown in Algorithm 13.5.

The working of this algorithm is depicted in Fig. 13.15. The matchings shown by different edge patterns denoting edge colors are $M_1 = \{e_1, e_9, e_4, e_6\}$, $M_2 = \{e_2, e_5, e_8, e_{13}\}$, $M_3 = \{e_3, e_{10}\}$, $M_4 = \{e_7, e_{14}\}$ resulting in one less color than the random example above. Time complexity of this method depends on the matching algorithm used.

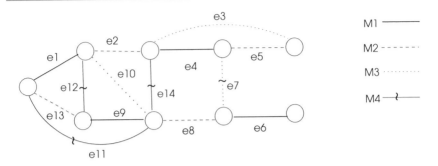

Fig. 13.15 Edge coloring using matching

Algorithm 13.5 *Ecolor_Match*

1: **Input** : $G = (V, E)$
2: **Output** : $\phi' : E \rightarrow C$ where $C = \{1, 2, ..., k\}$
3: $i \leftarrow 1$
4: **while** $E \neq \emptyset$ **do**
5:     **find** a maximal matching M_i of G
6:     **color** the edges of M_i with i
7:     $E \leftarrow E \setminus M_i$
8:     $i \leftarrow i + 1$
9: **end while**

13.6 Vertex Cover

A vertex cover of a graph G is a subset of its vertices such that any edge of G has at least one endpoint in the vertex cover set. We can have unweighted vertex cover of a graph with no weights associated with its vertices or weighted vertex cover when the vertices have weights.

13.6.1 Unweighted Vertex Cover

Definition 13.10 (*Unweighted vertex cover*) A *vertex cover* of an unweighted graph $G = (V, E)$ is a set $V' \subseteq V$ such that for any $(u, v) \in E$, either $u \in V'$, or $v \in V'$ or both are in V'. A *minimal vertex cover* (MVC) of a graph G is a vertex cover of G that does not contain any other vertex cover of G as a proper subset. A *minimum vertex cover* (MinVC) of G has the minimum cardinality among all vertex covers of G.

As in the case of a dominating set, we search a vertex cover that is minimal. A vertex cover that is not minimal is shown in Fig. 13.16a as bold vertices in which removal of the middle vertex from the cover still leaves the remaining vertices as the

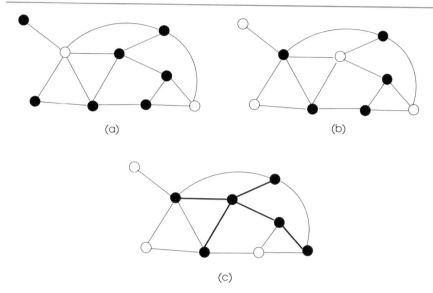

Fig. 13.16 a A MVC of a sample graph shown in bold **b** A MinVC of the same graph **c** A CVC of the same graph

vertex cover of the graph. The cover in (b) is minimal and also a minimum vertex cover of this graph. We can have a connected vertex cover (CVC) when there is a path between every pair of vertices in CVC that pass through only vertices in the vertex cover as shown in (c) with the path between cover vertices shown in bold.

An Approximation Algorithm

Finding a MVC V' of an unweighted graph $G = (V, E)$ can be done by finding edges of a matching M of G and including endpoints of all matched edges in the MVC as shown by the following algorithm steps.

1. **while** $E \neq \emptyset$
2.   Select an edge $(u, v) \in E$ at random.
3. $V' \leftarrow V' \cup \{u, v\}$.
4.   Remove (u, v) and all its adjacent edges from E.

The correctness of this algorithm is evident since we continue until no more edges left meaning every edge is covered by some vertex. Time complexity is $O(m)$ as in the greedy unweighted matching algorithm. Since two vertices are covered at each iteration, the approximation ratio of this algorithm is 2, that is, it finds a vertex cover that is at most twice the order of the MinVC. A vertex cover of a graph G is related to the independent set of G such that given an independent set I of G, $V - I$ is a vertex cover of G.

13.6.2 Weighted Vertex Cover

A weighted vertex cover (WVC) of a graph that has weights associated with its vertices still contains endpoints of every edge in the set, however, we aim to have a total minimum weight of vertices included in the cover as the main goal.

Pricing Algorithm

The pricing algorithm is designed to find the minimal weighted vertex cover (MWVC) of a vertex-weighted graph based on the following rules: an edge e when covered by a vertex pays a price p_e and the sum of the prices assigned to incident edges to a vertex v should not exceed the weight of v. The pricing algorithm first initializes the capacity of each vertex to its weight. An arbitrary unmarked edge (u, v) is selected at each step of the algorithm and if u or v has some capacity, edge (u, v) is charged with the lower of these capacities and whenever the capacity of a vertex is reduced to 0, it becomes *tight*, is included in the MWVC and all edges incident to that vertex are marked as covered. The procedure described favors vertices with lower weights by always selecting lower capacity endpoints of an edge to be tight and included in MWVC. Algorithm 13.6 displays the pseudocode of this algorithm. Note that any incident edge on a tight vertex is marked and excluded from further selections since a tight vertex is in MWVC.

Algorithm 13.6 *Pricing_MWVC*

1: Input $G(V, E, w)$ ▷ vertex weighted graph
2: $E' \leftarrow E, V' \leftarrow \emptyset$
3: **while** $E' \neq \emptyset$ **do**
4:     select any $(u, v) \in E'$
5:     **if** $c_u \neq 0 \vee c_v \neq 0$ **then**
6:         $w \leftarrow$ node with $\min(c_u, c_v)$
7:         $p(u, v) \leftarrow c_w, w \leftarrow tight$
8:         $V' \leftarrow V' \cup \{w\}$
9:         $E' \leftarrow E' \setminus \{(u, v)\} \cup$ any other edge incident at q
10:    **end if**
11: **end while**

Running of this algorithm in a sample graph with weighted vertices is shown in Fig. 13.17. The first selected edge is (a, f) and vertex f has the lower weight of 4 resulting in charging this edge with 4, reducing the capacity of vertex f to 0, including f in the vertex cover, and marking all edges incident to f as covered. Proceeding in this manner by arbitrarily selecting an uncovered edge results in the MWVC consisting of vertices $\{b, d, e, f\}$ with a total weight of 10.

Theorem 13.1 *Pricing Algorithm constructs a MWVC of a vertex-weighted graph with a time complexity $O(n)$ and an approximation ratio of 2.*

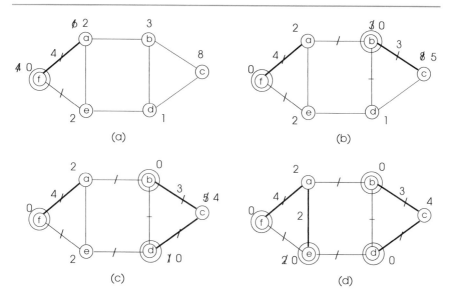

Fig. 13.17 Execution of Algorithm 13.6 in a sample graph

Proof Whenever a vertex is made tight, all edges incident to it are marked to be covered and the algorithm finishes when there are no more marked edges left. Thus, all edges are covered by tight vertices and the resulting tight vertex set is a vertex cover. Each step of the algorithm makes one vertex tight and there may be $O(n)$ steps in total to make at most n vertices tight.

Let V' be the set of all tight vertices produced by the algorithm and V_M the minimum vertex cover set that we are searching. We need to show $w(V') \leq 2w(V_M)$. The following can be stated when the total weights of vertices in V' and

$$w(V') = \sum_{v \in V'} w_v = \sum_{v \in V'} \sum_e p_e \leq \sum_{v \in V} \sum_e p_e \qquad (13.1)$$

Each edge is counted twice, therefore, Eq. 13.1 can be restated as follows.

$$w(V') = 2 \sum_{e \in E} \leq 2w(V_M) \qquad (13.2)$$

□

13.7   Review Questions

1. What is a maximum clique of a graph? Give an example.
2. What is the relationship between a clique and an independent set of a graph?
3. What is a matching of a graph?
4. Compare an independent set and a dominating set of a graph.

5. How is coloring performed on the vertices of a graph?
6. Compare vertex coloring and edge coloring of a graph.
7. What is the relationship between edge coloring and a matching of a graph?
8. What is a vertex cover of a graph and why do we search for a minimal vertex cover?

13.8 Chapter Notes

Some special subgraphs of a graph with a certain property have various applications. In this chapter, we reviewed few such subgraphs; cliques, edges of a matching, independent sets, dominating sets, vertex cover and also assigning colors to the vertices and edges of a graph. Some example applications of these subgraphs are as follows. Cliques may be used to find closely interacting nodes such as in a social network or a biological network, dominating sets may be used as a communication backbone and vertex cover may be used for facility placement as we will see in Chap. 15.

All of these problems except the matching problem are NP-hard which means no polynomial algorithm has been found to solve any of them. In such cases, we search for algorithms that use heuristics of some kind. The goodness of a used heuristic is commonly shown by extensive experiments on a wide range of data. Theoretically, approximation algorithms that find *proven* suboptimal solutions can be designed.

Exercises

1. State the vertex sets of the cliques in the graph of Fig. 13.18.
2. Find an MIS of the graph of Fig. 13.19 and check whether this is maximum.
3. Work out a MM in the graph of Fig. 13.20 and check whether an augmenting path in this matching exists.

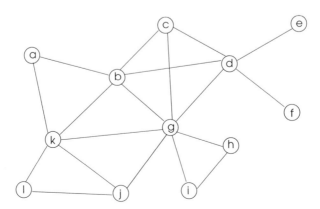

Fig. 13.18 Sample graph for Exercise 1

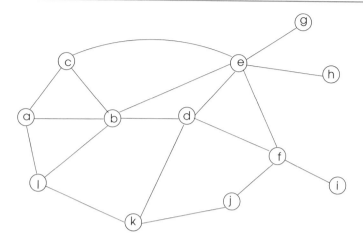

Fig. 13.19 Sample graph for Exercises 2 and 10

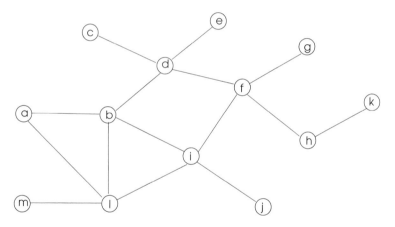

Fig. 13.20 Sample graph for Exercise 3

4. Show the steps of execution of Preis' algorithm to find MaxWM in the sample graph of Fig. 13.21.
5. The MaxIS of a tree T can be found by first including all leaves of T in the independent set and then not including the nodes in one level up and including all nodes in the next level up so on. Write this algorithm in pseudocode and work out its time complexity.
6. Show the running steps of Algorithm 13.2 (span-based algorithm) in the graph of Fig. 13.22.
7. Show that chromatic number of K_n is n.
8. Apply first the basic greedy and then the highest-degree-first vertex coloring to the graph of Fig. 13.23.

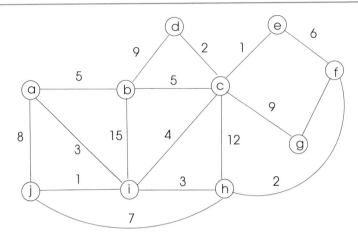

Fig. 13.21 Sample graph for Exercise 4

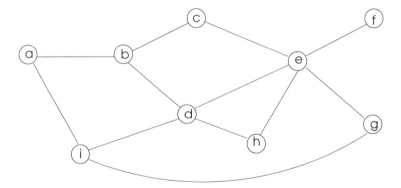

Fig. 13.22 Sample graph for Exercise 6

9. Prove that a tree can be colored using two colors only. Write the pseudocode of an algorithm to color the nodes of a tree and work out its time complexity.
10. Find disjoint matchings in the graph of Fig. 13.19 and provide an edge coloring of this graph based on these matchings.

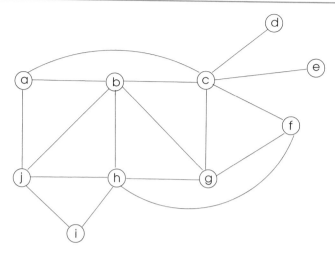

Fig. 13.23 Sample graph for Exercise 8

References

1. Berge C (1957) Two theorems in graph theory. Proc Natl Acad Sci USA 43:842–844
2. Bollobas B (1979) Graph theory. Springer, New York
3. Erciyes K (2018) Guide to graph algorithms: sequential, parallel and distributed. Springer texts in computer science series
4. Garey MR, Johnson DS (1978) Computers and intractability: a guide to the theory of NPcompleteness. Freeman, New York
5. Karp RM (1991) Probabilistic recurrence relations. In: Proceedings of the 23rd annual ACM symposium on theory of computing (STOC 91), pp 190–197
6. Preis R, (1999) Linear time 1, 2-approximation algorithm for maximum weighted matching in general graphs. In: Meinel C, Tison S (eds) Symposium on theoretical aspects of computer science (STACS), (1999) LNCS, vol 1563. Springer, Berlin, pp 259–269
7. Wattenhofer R (2016) Principles of distributed computing (Chapter 7). Class notes, ETH Zurich

Connectivity, Network Flows and Shortest Paths

14

Connectivity is a fundamental concept in graph theory as it has both theoretical and practical implications. A graph is called *connected* if it is possible to reach all vertices from any vertex. Connectivity is related to network flows and matching as we will see. In practice, connectivity is important in reliable communication networks as it has to be provided in loss of edges (links) or vertices (routers) in these networks.

We start this chapter by first reviewing the basic concepts related to connectivity and then describe algorithms to test connectivity of undirected and directed graphs. We provide a short review of network flows in the second part of the chapter and then conclude with matrix analysis of graphs related to connectivity.

14.1 Basics

A graph G is called *connected* if there is a path between every pair of its vertices; if G is not connected, it is called a *disconnected* graph. A graph may have subgraphs that are not connected to each other; these subgraphs are called *components*. Figure 14.1 displays a graph with 4 components; $C_1 = \{a, b, c, d, e\}$, $C_2 = \{f, g\}$, $C_3 = \{h\}$ and $C_4 = \{i, j, k\}$.

Definition 14.1 (*Vertex cut*) A *vertex-cut* of a connected graph $G = (V, E)$ is the set $V' \in V$ such that $G - V'$ has more components than G.

Note that $G - V'$ is the graph obtained when all vertices in V' with their incident edges are deleted from G. After this operation, the component number of G should increase. If V' has one vertex v only, v is called the *cut-vertex* of G. A graph with no cut-vertices is called a *bi-connected graph*. The largest component of the sample graph in Fig. 14.1 is bi-connected. The vertices $\{b, i\}$, $\{c, h\}$, $\{e, g\}$ are all different vertex cuts of the sample graph in Fig. 14.2 but vertex d is the only cut-vertex. Note

© Springer Nature Switzerland AG 2021
K. Erciyes, *Discrete Mathematics and Graph Theory*, Undergraduate Topics
in Computer Science, https://doi.org/10.1007/978-3-030-61115-6_14

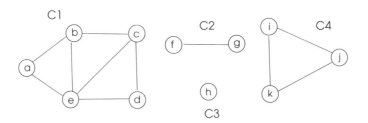

Fig. 14.1 Four components of a simple graph

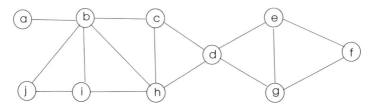

Fig. 14.2 Vertex cuts of a simple graph

that deletion of $V' = \{b, i\}$ results in three components of the graph whereas all other above stated removals result in two components.

Definition 14.2 (*Edge cut*) An *edge-cut* of a connected graph $G = (V, E)$ is the set $E' \in E$ such that $G - E'$ has at least two different components.

This time, we remove a set of edges from the graph G and test whether the number of components of G increases. When E' consists of a single edge, this edge is called a *bridge* of G. There are a number of edge cuts in the graph of Fig. 14.2 one example being $\{(c, d), (h, d)\}$ but (a, b) is the only bridge.

Definition 14.3 (*Vertex connectivity*) The *vertex-connectivity* denoted by $\kappa(G)$ of a connected graph G is the minimum number of vertices removal of which results in a disconnected or trivial graph.

When a graph G has a cut-vertex, this parameter is equal to 1 by definition. In general, it is desirable to have $\kappa(G)$ of a graph G that represents a network as large as possible since the network is vulnerable to failure of nodes. The range of values of vertex connectivity for any graph can be specified as below.

$$0 \leq \kappa(G) \leq n - 1 \tag{14.1}$$

The maximum value is $n - 1$ since even if every vertex of a graph is connected to every other vertex of the graph as in a complete graph K_n, we can remove $n - 1$ edges from any vertex to leave it isolated. A *k-connected* graph G has $\kappa(G) \geq 1$.

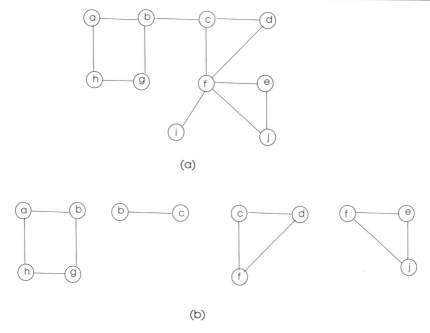

Fig. 14.3 **a** A sample graph **b** Some of its blocks

Definition 14.4 (*Edge connectivity*) The *edge-connectivity* $\lambda(G)$ of a connected graph G is the minimum number of edges removal of which results in a disconnected graph.

Using similar reasoning as in vertex connectivity, range of values of λ for a graph G can be defined as follows.

$$0 \le \lambda(G) \le n - 1 \tag{14.2}$$

Let $\delta(G)$ be the minimum degree present in a graph G. Then, tighter bounds on vertex and edge connectivity parameters can be imposed as below [3].

$$0 \le \kappa(G) \le \lambda(G) \le n - 1 \tag{14.3}$$

Definition 14.5 (*Block*) A *block* of a graph G is a maximal connected subgraph of G that does not contain a cut-vertex.

The blocks of a graph contains all bi-connected components, all bridges and all isolated vertices of the graph. Finding blocks of a graph shows us the parts of the graph that do not have cut-vertices. Some blocks of a sample graph are shown in Fig. 14.3.

14.1.1 Menger's Theorems

Given a graph $G = (V, E)$, let vertices $u, v \in V$. The *vertex connectivity* $\kappa(u, v)$ of these vertices is defined as the least number of vertices that are to be removed from the set V to leave u and v disconnected. Similarly, *edge connectivity* of two vertices $u, v \in V$, $\lambda(u, v)$, is the least number of vertices to be deleted from V to have u and v disconnected.

The *vertex disjoint paths* of two vertices u and v of a graph G are the paths that do not have any common vertices other than u and v. The *edge disjoint paths* of two vertices u and v of a graph G are the paths that do not have any common edges. It can be seen that the maximum number of vertex disjoint paths between u and v is $\kappa(u, v)$ and the maximum number of edge disjoint paths between u and v is $\lambda(u, v)$ since we need to remove as many vertices or edges to have these vertices disconnected. We can now state Menger's theorems which provide insight to the connectivity problem in graphs.

Theorem 14.1 (Menger's theorems) *Let $\kappa(u, v)$ be the maximum number of vertex disjoint paths between the vertices u and v of a graph $G = (V, E)$. A graph is k-connected if and only if each vertex pair in the graph is connected by at least k vertex disjoint paths. Let $\lambda(u, v)$ be the maximum number of edge disjoint paths between the vertices u and v of a graph $G = (V, E)$. Then G is k-edge-connected if and only if each vertex pair in the graph is connected by at least k edge-disjoint paths.*

14.2 Connectivity Test

A simple way to test whether an undirected simple graph G is connected or not is to run DFS or BFS algorithm in G starting from any vertex and record the visited vertices during the run in a list. If the list contains all of the vertices of G, then G is connected. This method works since each of these algorithms visits every connected vertices of G and the running time is the same for the DFS or BFS algorithm as $O(n + m)$.

A graph may not be connected consisting of a number of components. We may need to find the number of components and the vertices contained in each component of a graph G in this case. A simple modification to the basic recursive DFS algorithm provides the needed output as shown in Algorithm 14.1 [1]. Each call to the recursive procedure at line 11 signifies a new component and if all vertices of a graph are visited in a single call, then G is connected. The component identifier as a positive integer is stored in the array *Comp*, when all vertices have identifiers associated with integers $1, ..., n$, at the end of the algorithm. The time taken for this algorithm is $O(n + m)$ as the original DFS algorithm.

Algorithm 14.1 *DFS_Component*

1: **Input** : $G = (V, E)$
2: **Output** : $\mathcal{C} = \{C_1, C_2, ..., C_k\}$, $Comp[n]$: integer ▷ components of G and component numbers
   of each vertex
3: **boolean** $visited[1..n]$
4: $n\_comp \leftarrow 0$
5: **for all** $u \in V$ **do**   ▷ initialize
6:   $visited[u] \leftarrow false$
7: **end for**
8: **for all** $u \in V$ **do**
9:   **if** $visited[u] = false$ **then**
10:     $n\_comp \leftarrow n\_comp + 1$
11:     $DFS(u)$ ▷ call for each connected component
12:   **end if**
13: **end for**
14:
15: **procedure** $DFS(u)$
16:   $visited[u] \leftarrow true$
17:   $Comp[n\_comp] \leftarrow u$
18:   $C_{n\_comp} \leftarrow C_{n\_comp} \cup \{u\}$
19:   **for all** $(u, v) \in E$ **do**   ▷ recursively call neighbors
20:     **if** $visited[v] = false$ **then**
21:       $DFS(v)$
22:     **end if**
23:   **end for**
24: **end procedure**

14.3 Digraph Connectivity

The edges of a digraph have orientation and thus the connectivity of a digraph is defined differently than that of an undirected graph. A digraph $G = (V, E)$ is strongly connected if for each vertex pair $u, v \in V$, there exists a path from u to v and there exists path from v to u. In other words, we need connectivity in both directions since connectivity in one direction does not imply the other one.

14.3.1 Strong Connectivity Check

A procedure to test whether a digraph is strongly connected or not can be designed with the following idea. Starting from an arbitrary vertex v of the graph $G = (V, E)$, BFS or DFS algorithm is run and the visited vertices are noted. Then the transpose of G, G^T, is obtained by reversing the direction of the edges of G and the BFS or DFS algorithm is executed again from vertex v. If the visited vertices in both cases is equal to V, then the digraph is strongly connected. This method is illustrated as a procedure in Algorithm 14.2.

Algorithm 14.2 *Strong_Conn*

1: **procedure** TEST_SCC($G = (V, E)$)   ▷ return *true* if G is strongly connected else return *false*
2:     $X \leftarrow \emptyset, Y \leftarrow \emptyset,$
3:     **choose** an arbitrary vertex $v \in V$
4:     $DFS(G, v)$
5:     **save** the visited vertices in X
6:     $G^T \leftarrow G$ with edges reversed
7:     $DFS(G^T, v)$
8:     **save** the visited vertices in Y
9:     **if** $X = Y = V$ **then**
10:         **return** *true*
11:     **else**
12:         **return** *false*
13:     **end if**
14: **end procedure**

An example digraph is depicted in Fig. 14.4a and DFS algorithm is run from vertex v_8 with the formed tree lines shown in bold and first visit times of vertices are shown next to them. All of the vertices are reached with this running and then G_T in (b) is obtained by reversing the directions of edges of the digraph in (a). DFS algorithm is now executed from vertex v_8 in G_T and all vertices are again accessed which means G is strongly connected. A change of edge directions may result in a not strongly connected digraph, for example, if $(v_3, v_4) \in E$ in (a) instead of $(v_4, v_3) \in E$, G would not be strongly connected.

A digraph may not be strongly connected but may have strongly connected subgraphs. Such a component has paths between each pair of vertices it contains. A digraph with 3 strongly connected components C_1, C_2 and C_3 is shown in Fig. 14.5.

14.3.2 Finding Strongly Connected Components

Finding strongly connected components (SCCs) of a digraph G which are subgraphs of G that are strongly connected has various applications such as detecting closely interacting communities in a social network. The SCCs of a digraph G are the same in the transpose of G which is obtained by reversing the edge directions in G. Also, the contraction of a digraph G into G^{SCC} shrinks the vertices of G into supervertices of G^{SCC} each of which is a SCC of G. The contraction is performed such that for any supervertex pair U and V, if there is an edge between $u \in U$ and $v \in V$ in G, then there exists an edge between U and V in G^{SCC}. An algorithm based on the following observations which we will state without proof may be formed.

- In any DFS, all vertices of the same SCC are in the same DFS tree.
- For any two vertices u and v that are in the same SCC, no path between them leaves the SCC.
- For any directed graph, G^{SCC} is an acyclic digraph.

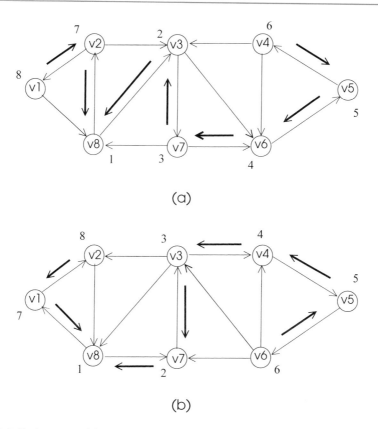

Fig. 14.4 Testing connectivity of a digraph

The algorithm due to Kosaraju is based on the above observations and works with the following steps.

1. **Input**: A digraph $G = (V, E)$
2. **Output**: SCCs of G
3. Perform DFS on G and make a list L of vertices in the non-increasing order of their finishing times.
4. Construct G^T by reversing the direction of each edge of G.
5. Perform DFS on G^T starting from the front of L.
6. Return DFS trees obtained this way.

An example digraph with SCCs is depicted in Fig. 14.6. Running DFS in this graph from vertex d results in the first and last visit times of the vertices shown next to them. The list L shown is then formed by sorting the last visit times of vertices.

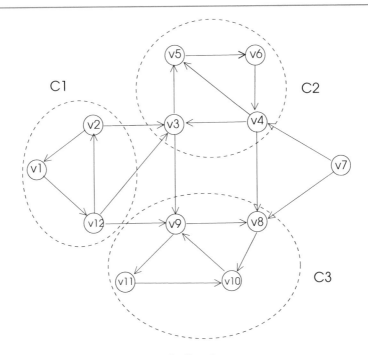

Fig. 14.5 Strongly connected components of a digraph

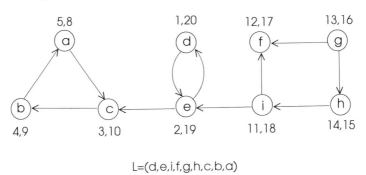

L=(d,e,i,f,g,h,c,b,a)

Fig. 14.6 Strongly connected components of a digraph

We form G^T by reversing edge directions to obtain the graph of Fig. 14.7. Running DFS starting by the first vertex d from the list L on G^T results in the DFS trees T_1, T_2 and T_3 each of which is a SCC of G as depicted in this figure. The running time of this algorithm is running DFS twice which is $O(n + m)$.

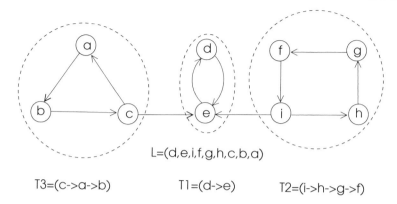

L=(d,e,i,f,g,h,c,b,a)

T3=(c->a->b)   T1=(d->e)   T2=(i->h->g->f)

Fig. 14.7 Strongly connected components of a digraph

14.4 Network Flows

Let $G = (V, E)$ be a digraph with two special nodes, a start node s that has only outgoing edges and a sink node t having only ingoing edges. Each edge e of the digraph G is labeled with a non-negative capacity $c(e)$ defined by $c : E \to \mathbb{R}^+$. The *network flow problem* is to provide as much flow as possible from s to t without exceeding the capacity of edges and the total flow into a node should equal to the total flow out from the node. Formally, the following rules are applied with $f(e)$ denoting the flow through an edge e,

- *Capacity limit*: $\forall e \in E$, $f(e) \leq c(e)$
- *Conservation of flow*: $\forall v \in V\{s, t\}$, $\sum_v f(u, v) = \sum_v f(v, u)$

A flow f on a network is the function $f : E \to \mathbb{R}^+$. A flow network is shown in Fig. 14.8 with the node s having only flow out and the node t having only flow in. Three different flows from s to t are shown in dashed lines and the total flow value is the sum of these flows which is 8. Note that this is the maximum flow through this network as the node s can not put more flow to the network due to the saturation of the capacities in its outgoing edges.

Definition 14.6 (*s-t cut, cut value*) An *s-t cut* (A, B) in a digraph $G = (V, E)$ partitions the vertices of G into subsets A and B such that $s \in A$ and $t \in B$. The *cut-value* of a cut (A, B) is the sum of the capacities of all edges with direction from A to B.

Fig. 14.8 Flows through a
network

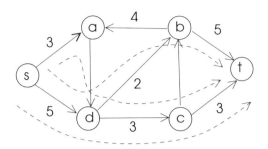

14.4.1   A Greedy Algorithm

Based on what we saw so far, we can have a greedy algorithm with the following
idea. Starting with a zero flow, we gradually increase flow value at each iteration by
selecting an $s - t$ path, finding the minimum possible improvement edge along this
path and increase the flow value by increasing all edge flows by this amount.

1. **Input**: $G = (V, E)$
2. **Output**: Maximum flow f from s to t in G
3. **while** there is such a path **do**
4.   **find** an $s - t$ path P such that $\forall e \in P, f(e) < c(e)$
5. $d \leftarrow min_{e \in P}(c(e) - f(e))$
6.   **for all** $e \in P$ **do**
7. $f(e) \leftarrow f(e) + d$
8.   **end for**
9. **end while**

Let us implement this algorithm in the graph of Fig. 14.8 and we can see it will
stop when there is not an *s-t* path with an edge that has flow less than its capacity.
This greedy algorithm will not provide the optimal solution in general and further
optimizations are needed.

14.4.2   Residual Graphs

The residual graph G_f of a digraph $G = (V, E)$ representing a network is defined
by two rules:

- $\forall (u, v) \in E, c_f(u, v) = c(u, v) - f(u, v)$
- if $f(u, v) > 0$ then $c_f(v, u) = f(u, v)$

Essentially, G_f shows us where to increase or decrease flows.

Definition 14.7 (*Augmenting path*) An augmenting path is a directed path from the
node s to the node t in the residual graph G_f.

Existence of an augmenting path in a residual network means we can push more flow through this path in the original network. Let $(s, v_1, v_2, ..., v_k, t)$ be an augmenting path in the graph G_f. The maximum flow that can be pushed through this path is $min(c(s, v_1), c(v_1, v_2), ..., c(v_k, t))$. Thus, given a graph G of a network, if there exists an augmenting path in G_f, then the flow f is not maximum.

14.4.3 Ford–Fulkerson Algorithm

Ford-Fulkerson algorithm makes use of the fact that whenever there is an augmenting path in a residual graph of a network, the total flow f can be increased by the minimum currently available capacity along the augmenting path. The augmenting paths are searched iteratively and the total flow f is increased using this principle as shown below by first initializing the residual graph G_f with the original graph G. The reverse flows between nodes of the graph are also modified such that a flow in reverse direction is decreased by the minimum capacity c found along the augmenting path. If there is no such reverse edge, a new edge is created with the flow c. This operation is needed since we may need to send flow through these edges to push more flow from s to t in the next iterations.

1. **Input**: $G = (V, E)$
2. **Output**: Maximum flow f from s to t in G
3. $G_f \leftarrow G$
4. **while** there is such a path **do**
5.   **find** an augmenting path P in G_f
6. $d \leftarrow min_{e \in P}(c(e) - f(e))$
7.   **for all** $e \in P$ from s to t **do**
8. $f(e) \leftarrow f(e) + d$
9.   **end for**
10.   **for all** $e \in P$ that has a corresponding reverse edge in P **do**
11. $f(e) \leftarrow f(e) - d$
12.   **end for**
13. **end while**

Key to the operation of this algorithm is searching for augmenting paths in the residual graph G_f and also, the capacities of the edges of G_f need to be modified at each step. The iterations of Ford–Fulkerson algorithm in a small flow network is shown in Fig. 14.9. Each newly found augmenting path shown by bold edges increases the flow by the minimum current capacity edge through the path and the maximum flow found this way is 6 and the final flow network is shown in (e). When the flow values are integers, we can increase the flow value by $|f^*|$ times where f^* is the maximum flow that can be attained in the network, assuming flow is increased at least once in each step. The search for an augmenting path can be performed in $O(n + m)$ time by a modified DFS algorithm, thus, the total time taken by this algorithm is $O(|f^*|(n + m))$.

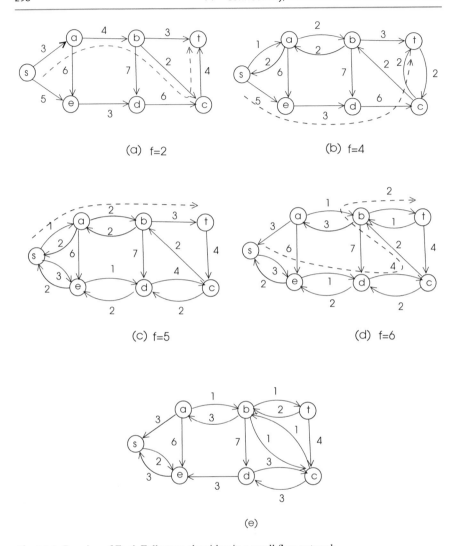

Fig. 14.9 Running of Ford–Fulkerson algorithm in a small flow network

14.4.4 Bipartite Graph Matching

We reviewed bipartite graphs and the matching problem in these graphs in Sect. 13.2.3
while searching for a maximal number of edges that are not adjacent to each other.
Let $G = (A \cup B, E)$ be an unweighted bipartite graph with A the set of vertices on
the left and B the set of vertices on the right of the graph. A matching $M \subseteq E$ is
a maximum matching if it has the largest size among all matchings of G. Matching
problem in a bipartite graph can be formulated as a network flow problem by applying
the following steps.

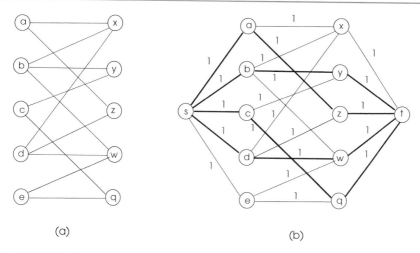

Fig. 14.10 Transforming a bipartite graph to a flow network

1. Obtain the graph $G' = (V', E')$ from G.
2. Find the maximum flow f in G' using an algorithm like Ford–Fulkerson.
3. Use f to construct a maximum matching M in G.

These steps will be clearer as we proceed. The first step is achieved by the following sub-steps.

1. Add a source node s and a sink node t to G.
2. For each edge (u, v) from s to t, add direction from u to v and label each edge with capacity 1.
3. Add an edge with capacity 1 to each node of A from node s.
4. Add an edge with capacity 1 to node t from each node of B.

This transformation of a bipartite graph to a flow network is depicted in Fig. 14.10. The next step is to find the maximum flow in this flow network, say using Ford–Fulkerson algorithm. The flow is incremented at each iteration and there are 4 augmenting paths in Fig. 14.10b shown in bold resulting in a total flow of 4. Therefore the size of maximum matching is 4 and the matched edges are shown in bold excluding the edges between the node s and nodes in set A, and the edges between the nodes in set B and the node t.

14.5 Algebraic Connectivity

We have already reviewed adjacency matrix and incidence matrix of a graph which show the structure of a graph. We now need to define new metrics for graph connectivity.

14.5.1 The Laplacian Matrix

The degree matrix of a graph $G = (V, E)$ is a diagonal matrix $D(G)$ of G which has $D[i, i]$ element as the degree of vertex i, and all other elements of D are 0s.

The Laplacian matrix $L(G)$ of a graph G is defined as the difference of the diagonal matrix and its adjacency matrix as follows.

$$L(G) = D(G) - A(G) \qquad (14.4)$$

Thus, $L(G)$ has the degree values of vertices at its diagonal and $L[i, j] = -1$ if edge $(i, j) \in E$ and 0 otherwise as shown below.

$$l_{ij} = \begin{cases} d_i & \text{if } i = j \\ -1 & \text{if } i \text{ and } j \text{ are neighbors} \\ 0 & \text{otherwise} \end{cases}$$

Laplacian matrix of the graph in Fig. 14.11 is as follows.

$$
L = \begin{array}{c} \\ v_1 \\ v_2 \\ v_3 \\ v_4 \\ v_5 \\ v_6 \end{array}
\begin{array}{c} \begin{matrix} v_1 & v_2 & v_3 & v_4 & v_5 & v_6 \end{matrix} \\
\left(\begin{matrix}
2 & -1 & 0 & 0 & 0 & -1 \\
0 & 4 & -1 & 0 & -1 & -1 \\
0 & -1 & 3 & -1 & -1 & 0 \\
0 & 0 & -1 & 2 & -1 & 0 \\
0 & -1 & -1 & -1 & 4 & -1 \\
-1 & -1 & 0 & 0 & -1 & 3
\end{matrix} \right) \end{array}
$$

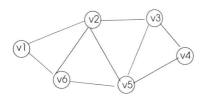

$$
\begin{array}{c} \\ v_1 \\ v_2 \\ v_3 \\ v_4 \\ v_5 \\ v_6 \end{array}
\begin{array}{c} \begin{matrix} v_1 & v_2 & v_3 & v_4 & v_5 & v_6 \end{matrix} \\
\left(\begin{matrix}
2 & 0 & 0 & 0 & 0 & 0 \\
0 & 4 & 0 & 0 & 0 & 0 \\
0 & 0 & 3 & 0 & 0 & 0 \\
0 & 0 & 0 & 2 & 0 & 0 \\
0 & 0 & 0 & 0 & 4 & 0 \\
0 & 0 & 0 & 0 & 0 & 3
\end{matrix} \right) \end{array}
$$

Fig. 14.11 A sample graph and its degree matrix

14.5.2 Normalized Laplacian

The *normalized Laplacian* matrix is defined as follows.

$$\mathcal{L} = D^{-1/2} L D^{-1/2} = D^{-1/2}(D - A)D^{-1/2} = I - D^{-1/2} A D^{-1/2} \qquad (14.5)$$

The entries of the normalized Laplacian matrix $\mathcal{L}$ can then be specified as below.

$$\mathcal{L}_{ij} = \begin{cases} 1 & \text{if } i = j \\ \dfrac{-1}{\sqrt{d_i d_j}} & \text{if } i \text{ and } j \text{ are neighbors} \\ 0 & \text{otherwise} \end{cases}$$

14.5.3 Eigenvalues

Let A be an $n \times n$ matrix and x a vector of size n. The product $A \times x$ results in a vector y and this operation can be written as, $Ax = y$. If we can have $y = \lambda x$ for some constant λ, this constant is called an *eigenvalue* of A and the vector x is called an *eigenvector* of A. There will be more than one eigenvalue and a set of eigenvectors of matrix A in general. The following can be derived from the definition of eigenvalue and eigenvector with I as the identity matrix having all 1s in its diagonal and 0 for all other elements.

$$Ax - \lambda x = 0$$
$$(A - \lambda I)x = 0$$
$$det(A - \lambda I)x = 0$$

The determinant of the last statement is called the *characteristic polynomial* of matrix A. It has n roots which are the eigenvalues of matrix A when the size of A is $n \times n$. Finding eigenvalues provides us with the eigenvectors by way of substitution in the equation $Ax = \lambda x$.

We are now ready to state the relationship between graph connectivity, Laplacian matrix and eigenvalues. We can evaluate the eigenvalues of the Laplacian matrix L and the following results become available [2]:

- All eigenvalues of L are positive except the smallest one which is 0.
- The number of eigenvalues of L with the value 0 is equal to the number of components of the graph that L represents.
- The graph G shown by L is connected if second smallest eigenvalue of L is positive. The larger this value, the more connected G is.

The second eigenvalue of the Laplacian matrix shown by $\sigma(G)$ is denoted by the *Fiedler value* of the graph G. The set of all eigenvalues of the Laplacian matrix of a graph G is called the *Laplacian spectrum* of G.

14.6 Shortest Paths

We saw how BFS can be used to find shortest paths from a given node in an un-weighted graph. The minimum number of hops to the source vertex was considered as the distance to it. This in fact is equivalent to the case of a weighted graph with each edge having unity weight. In the more general case when edges of a graph may have arbitrary weights, different procedures are needed.

The *length* of a path $p =< v_0, v_1, ..., v_k >$ in a weighted graph $G = (V, E, w)$ is the sum of the weights of edges included in the path as shown below.

$$length(p) = \sum_{i=1}^{k} w(v_{i-1}, v_i)$$

Distance from a vertex u to a vertex v in a weighted graph is the length of the minimum length path if such a path exists and is ∞ otherwise. The single source shortest path (SSSP) problem is finding distances from a source vertex to all other vertices in a weighted directed or undirected graph.

Dijkstra's SSSP algorithm aims to find the shortest paths from a source vertex to all other vertices in a weighted directed or undirected graph with positive edge weights. It is an iterative algorithm that processes a vertex in each step. The idea is similar to Prim's MST algorithm that selects the minimum weight outgoing edge at each iteration, this time however, the vertex that has the shortest path to the source vertex is selected at each iteration. Key to the operation of this algorithm is the updating of neighbor vertex distances when a vertex is processed. The pseudocode of this algorithm is shown as a procedure in Algorithm 14.3 which inputs a weighted undirected or directed graph, first initializes distances of neighbors of source vertex s. It then finds minimum distance vertex v, includes in the shortest path tree and updates distances of u at each iteration.

Running of this algorithm in a digraph with 7 nodes is depicted in Fig. 14.12 with current distance values of vertices shown in italic next to them and the distance selected vertex is shown in bold. Starting by the first iteration in (a), v_7 has the minimum label and is included in the tree T to be formed. The weights of its neighbor v_2 is changed to 4 from the initial 8 value, v_3 from ∞ to 3, and v_6 from ∞ to 13. The minimum label vertex in (a) is v_3 and is selected to be in T and its neighbor distances are updated in (b). This procedure continues until all vertices are processed and the spanning tree shown by bold lines is formed in (e) with each vertex except the root pointing to its parent in this tree by bold arrows.

The straight forward running time of this algorithm is $O(nm)$ since the outer loop of the algorithm runs at most $n - 1$ times considering each vertex and the inner loop examines each edge in at most m time. Almost linear-time implementations may be achieved using data structures such as heaps and priority queues [1].

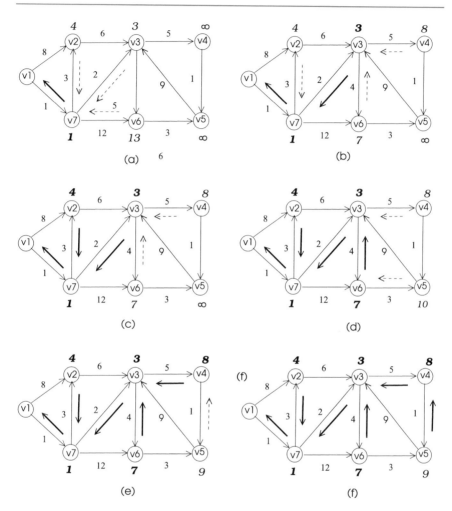

Fig. 14.12 Execution of Dijkstra's SSSP algorithm in a sample digraph from vertex v_1

Algorithm 14.3 *Dijkstra_SSSP*

1: **procedure** DIJKSTRA(G,s)
2:    **Input** : $G(V, E, w)$ ▷ connected, weighted graph G and a source vertex s
3:    **Output** : $D[n]$ and $P[n]$ ▷ distances and predecessors of vertices in the tree
4:              tree vertices T
5:    **for all** $u \in V \setminus \{s\}$ **do**   ▷ initialize all vertices except source s
6:        $d(u) \leftarrow \infty$
7:    **end for**
8:    $d[s] \leftarrow 0$; $pred(s) \leftarrow s$
9:    $V' \leftarrow \emptyset$; $T \leftarrow \emptyset$
10:   **while** $V' \neq V$ **do**
11:       **select** $u \notin V'$ with minimum $d(u)$
12:       $V' \leftarrow V' \cup \{u\}$
13:       **for all** $(u, v) \in E$ **do**   ▷ update neighbor distances to u
14:           **if** $d(v) > d(u) + w(u, v)$ **then**
15:               $d(v) \leftarrow d(u) + w(u, v)$
16:               $pred(v) \leftarrow u$ ▷ update tree structure
17:           **end if**
18:       **end for**
19:       $T \leftarrow T \cup \{u, v\}$ ▷ add it to tree vertices
20:   **end while**
21: **end procedure**

14.7   Chapter Notes

We reviewed graph connectivity and related concepts in this chapter. Investigation of the connectivity of a graph provides vital information about the structure of the graph which can be used for many applications. A computer network should be connected at all times and finding vulnerable regions of a network may provide where to strengthen the network.

We started with basic definitions and descriptions of concepts such as vertex and edge connectivity of a graph and moved on to algorithms to test whether a graph is connected or not. We saw simple modifications of BFS or DFS algorithms provide this information. We need to check orientation of edges when connectivity in digraphs is considered since paths between vertex pairs are bidirectional. If A is the adjacency matrix of an undirected or a directed graph, the nth power of A, A^n, has an entry (i, j) which is equal to number of walks between the vertices i and j.

Connectivity is closely related to network flow concept and we reviewed the relationship between these two seemingly diverse areas of study. Network flow may be used in transportation systems for the transportation of goods, manufacturing systems for the flow of items, and in communication systems for flow of data across the networks. The algebraic properties of a graph can be used to find how connected it is by analyzing the eigenvalues of the Laplacian matrix. Lastly, we investigated finding shortest paths between vertices in weighted graphs and reviewed a well-

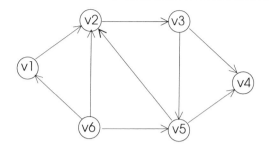

Fig. 14.13 Example graph for Exercise 4

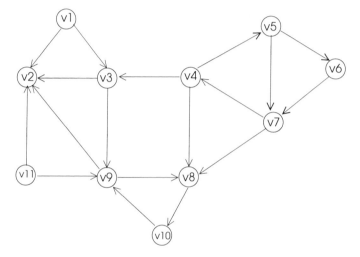

Fig. 14.14 Example graph for Exercise 5

Fig. 14.15 Example graph
for Exercise 6

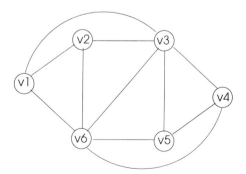

known algorithm due to Dijkstra for this purpose. The area of connectivity is vast
and a comprehensive survey of the topic can be found in [1].

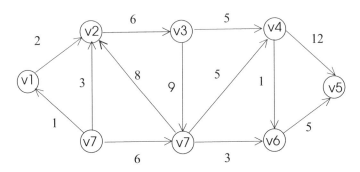

Fig. 14.16 Example graph for Exercise 7

Exercises

1. Show that for every graph,

$$\kappa(G) \le \lambda(G) \le \Delta(G)$$

2. A simple way to find the cut-vertex of a graph G is to remove vertices one by one and check whether the remaining graph is connected or not by the DFS or the BFS algorithm. Note that we need to place a removed vertex back to the graph if it is not a cut-vertex before the next iteration. Write the pseudocode of this algorithm and work out its time complexity.

3. Write the pseudocode of an algorithm that finds whether an edge of a graph G is a bridge or not by removing that edge and checking for connectedness of G and find the time complexity of this algorithm.

4. Test whether the digraph of Fig. 14.13 is strongly connected or not.

5. Use Kosaraju's algorithm to find the strongly connected components of the digraph shown in Fig. 14.14.

6. Find the Laplacian matrix of the graph shown in Fig. 14.15.

7. Work out the distances to vertex v_1 of the weighted digraph of Fig. 14.16.

References

1. Erciyes K (2018) Guide to graph algorithms: sequential, parallel and distributed. Springer texts in computer science series
2. Fiedler M (1989) Laplacian of graphs and algebraic connectivity. Comb Graph Theory 25:57–70
3. Whitney H (1932) Congruent graphs and the connectivity of graphs. Am J Math 54:150–168

Graph Applications

15

Graphs whether undirected or directed, weighted or unweighted have numerous applications. A graph can be used to represent a network of any kind as we have seen which means any network application can make use of graph theory. We will review such graph application areas centered around networks in this chapter, starting with the analysis of large graphs. These networks are computer networks, ad hoc wireless networks, biological networks and social networks.

15.1 Analysis of Large Graphs

Many real-life networks are large consisting of thousands of nodes and tens of thousands of edges. For example, Internet is one such network with billions of nodes and various social networks have billions of users. Analyzing such large networks is a challenge, we can not visualize them in complete but we need to investigate them to understand their structure and functioning. We first need o define some parameters to analyze large networks.

15.1.1 Degree Distribution

The *degree distribution* $P(k)$ in an undirected graph is the fraction of vertices in the graph with degree k defined as follows.

$$P(k) = \frac{n_k}{n},$$

where n_k is the number of vertices with degree k and n is the number of vertices. We can plot $P(k)$ against the degrees of a graph to inspect the degree distribution of a graph visually. The degree distribution of a random network is binomial in the shape of a bell called *bell curve* or *normal distribution*. Degree distribution of a

© Springer Nature Switzerland AG 2021
K. Erciyes, *Discrete Mathematics and Graph Theory*, Undergraduate Topics
in Computer Science, https://doi.org/10.1007/978-3-030-61115-6_15

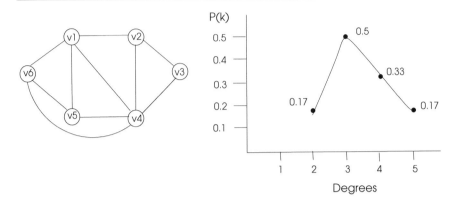

Fig. 15.1 Degree distribution of a sample graph

sample graph is shown in Fig. 15.1 with $P(2) = 1/6 = 0.17$, $P(3) = 3/6 = 0.5$, $P(4) = 1/3 = 0.33$, $P(5) = 1/6 = 0.17$ values.

15.1.2 Clustering

Finding a subgraph consisting of closely related elements in a graph has numerous applications such as detecting a group of friends in a social network. *Clustering* is the process of grouping similar nodes of a graph or any other structure. Clustering is widely studied in various disciplines including Computer Science, Statistics and Bioinformatics with numerous algorithms proposed for this purpose. Clustering proteins inside the cell in a biological network may provide insight to health and disease conditions.

We will review a simple and effective clustering algorithm based on the minimum spanning tree (MST) discussed in Chap. 12. An MST of a weighted graph $G = (V, E, w)$ is a spanning tree of G that has a total minimum total edge weights among all spanning trees of G as noted. Let us assume that we have a weighted graph G that represents a network with weights inversely proportional to some relationship among nodes of the network, thus, our aim is to group nodes that have as much light edges as possible among them. Constructing an MST in such a graph performs the first elimination by discarding the heavy edges initially. The MST based clustering method works as follows.

1. **Input**: $G = (V, E, w)$ ▷ a weighted graph
2. **Output**: $C = \{C_1, C_2, .., C_k\}$ ▷ a set of clusters of G
3. **construct** an MST T of G using Prim's, Kruskal's or other algorithm
4. **while** the number of required clusters not reached
5.   **remove** the heaviest edge (u, v) from T
6.   **form** the new clusters by performing BFS from u and from v
7. **endwhile**

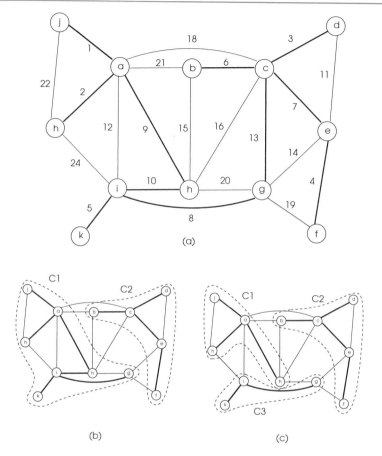

Fig. 15.2 MST-based clustering

Removing the heaviest edge at each iteration requires sorting the edges of T which is $O(m \log m)$ time. Forming clusters means we need to include node labels in the clusters formed. A simple way to insert nodes into new clusters is to run BFS from each endpoint of the removed edge (u, v). Total time taken is $O(m \log m + n + m)$ in this case. Running of this algorithm is depicted in Fig. 15.2 where an MST of the graph is shown in bold edges in (a). We remove the heaviest edge (c, g) from T first to result in two clusters C_1 and C_2 a shown in (b). The second iteration of the algorithm breaks cluster C_1 into clusters C_1 and C_3 by removing the current heaviest edge (i, h) from the MST T as in (c). Thus, we need to run the while loop $k - 1$ times to have k clusters. More generally, we may not know the number of clusters to find initially. In such a case, we need to stop on a condition such as the weight of the next edge to be removed is lower than some pre-determined threshold.

We need to define some parameters to be able to asses the clustering structures of large graphs.

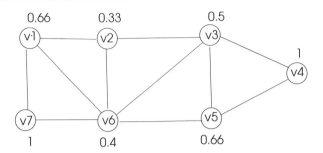

Fig. 15.3 Clustering coefficients of vertices of a graph

Definition 15.1 (*Clustering coefficient*) The *clustering coefficient* of a vertex v in a graph $G = (V, E)$ is the ratio of the existing edges between the neighbors of vertex v to all possible number of edges between these neighbors. Formally,

$$cc(v) = \frac{2m_v}{n_v(n_v - 1)}$$

where n_v is the number of neighbors of vertex v, and m_v is the existing number of edges between these neighbors.

Note that $n_v(n_v - 1)/2$ is the maximum possible connections between these neighbors. This parameter shows how closely the neighbors of a vertex are connected. The average clustering coefficient $CC(G)$ of a graph G is the arithmetic average of all the clustering coefficients of vertices in G. Formally,

$$CC(G) = \frac{1}{n} \sum_{v \in V} cc(v)$$

The average clustering coefficient gives us some idea on how dense the graph is connected. Figure 15.3 displays clustering coefficients of vertices in a sample graph. The average clustering coefficient of this graph is 0.65.

15.1.3 Matching Index

Matching index is a parameter used to asses the similarity between two vertices in a graph. Matching index of two vertices u and v in a graph G is defined as the ratio of the number of common neighbors of u and v to the sum of their neighbors. In formula, matching index MI of u and v is as follows,

$$MI(u, v) = \frac{n_{uv}}{n_u + n_v}$$

where n_{uv} is the number of common neighbors and n_u and n_v are the number of neighbors of u and v respectively. Matching indices of vertex pairs of a sample graph is shown in Fig. 15.4. Dashed lines are used to display matching index between two

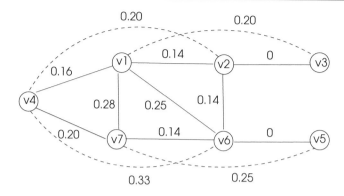

Fig. 15.4 Matching indices in a sample graph

vertices and a matching index of two vertices with no common neighbors is omitted. We can see $MI(v_4, v_6)$ has the largest value, thus, these two vertices match best in this graph.

15.1.4 Centrality

The importance of a vertex or an edge in a large graph can be assessed by evaluating the *centrality* parameter. The basic centrality of a vertex is simply its degree and using this parameter, we can state that a higher degree vertex is more important than a lower degree vertex, for example, a person in a social network with many friends may be considered important in terms of her connections. The average degree of a network gives us some idea about the structure of the graph representing a network but it is difficult to have an overall estimate of the graph structure based on this parameter only.

15.1.4.1 Closeness Centrality
The closeness centrality of a vertex is another centrality parameter to asses the easiness of reaching all other vertices from that vertex.

Definition 15.2 The *closeness centrality* $C_C(v)$ of a vertex v is the reciprocal of the sum of its distances to all other vertices in the graph given as below,

$$C_C(v) = \frac{1}{\sum_{u \in V} d(u, v)} \tag{15.1}$$

Finding distances from a vertex to all other vertices in an unweighted graph can be performed by using the BFS algorithm. In a weighted graph, Dijkstra's single source shortest path algorithm may be used. Closeness centralities of the vertices of a simple graph is calculated as shown next to the vertices in Fig. 15.5. We can see that vertex v_1 has the highest centrality as it can reach all other vertices more easily than others.

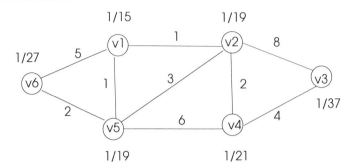

Fig. 15.5 Closeness centralities in a sample graph

15.1.4.2 Betweenness Centrality

In yet another approach to asses the significance of a vertex or an edge in a graph is to evaluate the percentage of shortest paths that run through a vertex or an edge in a graph. Vertex betweenness is used to evaluate the former and edge betweeness is the assessment of the latter.

Definition 15.3 (*Vertex betweenness*) *Vertex betweenness* of a vertex v in a graph is the sum of the ratios of the number of shortest paths between every vertex pair (s, t) that run through vertex v to the sum of any shortest path between vertex pair (s, t) as follows.

$$V_C(v) = \sum_{s \neq t \neq v} \frac{\sigma_{st}(v)}{\sigma_{st}} \tag{15.2}$$

where $\sigma_{st}(v)$ is the total number of shortest paths that pass through vertex v and σ_{st} is the number of any shortest paths that go through s and t. Informally, this parameter provides an importance of a vertex in a graph, the more shortest paths run through it, the more important it is.

Edge betweenness is defined similarly to vertex betweenness, this time we search shortest paths that run through a specific edge e instead of a vertex as shown by the equation below.

$$B(e) = \sum_{s \neq t \neq v} \frac{\sigma_{st}(e)}{\sigma_{st}} \tag{15.3}$$

15.1.5 Network Models

Large networks are mainly classified as random networks, small world networks and scale-free networks.

Random Networks

The random network model (ER-Model) proposed by Erdos and Renyi [6] assumes that the probability to have an edge between any pair of nodes is distributed uniformly at random. In any graph $G = (V, E)$, the maximum number of possible edges is $n(n-1)/2$. Each edge is added to the network with the probability $p = 2m/n(n-1)$ in this model. However, ER-Model does not reflect most real networks and new models are developed.

Small-World Networks

In many real networks such as the Internet and biological networks, it is often the case that distances between the nodes of such networks are small. These networks are called *small-world* networks. A social experiment performed by the sociologist Milgram aimed at determining the probability that two randomly selected people would know each other [11]. Individuals of certain U.S. cities were selected as the starting points and some other cities as the target cities. Letters were sent to the individuals with some names included in them and they were asked to send the letters directly to the person in the letter if they knew the person or to a person who would likely know the target person. The average number of intermediate people for most of the letters to reach the destinations was six and hence these experiments were associated with the *six degrees of separation* phrase. The diameter of a small-world network increases with the logarithm of the network size, formally, $d \approx \log n$ as $n \to \infty$ in these networks.

Scale-Free Networks

Many real networks exhibit few very high-degree vertices and many low-degree vertices. A social network such as a class of students may have few very popular students who are friends with many other students in class but most of the students will have only few friends in general. These networks are characterized by the power law degree distribution,

$$p(k) \sim k^{-\gamma} \quad \text{with } \gamma > 1$$

which basically means that the probability of the existence of a node gets smaller when the node degree k increases. These networks are termed *scale-free networks* based on the work of Barabasi and Albert [1].

15.2 The Web

A computer network consists of a collection of computational nodes commonly called *hosts* connected by various communication media such as copper wire, fiber, wireless medium and various communication components. A main component in a computer network is the *router* which is used to switch any incoming message to one of its output channels. Modern communication networks provide communication in both directions, thus, an undirected graph is commonly used to represent such networks. The cost of sending a message from one node of the network to another depends on factors such as the structure of the physical medium, speed of nodes etc. and thus can not be assumed same for all connections. Therefore, we need to use a weighted graph to represent a computer network with weights over the edges of the graph displaying the cost of sending a message between the two endpoints of the edge.

Routing performed by a router is the process of switching an incoming message to one of the output channels this component has. Let us assume that each router is aware of the network topology, in other words, the structure of the graph representing the computer network. Single-source Shortest-path (SSSP) algorithms find all possible routes to transfer a message from a source node to all other nodes in the network. We have two classical algorithms for this purpose; Dijkstra's SSSP algorithm and Bellman-Ford SSSP algorithm and contemporary computer networks use variations of these algorithms.

15.2.1   The Web Graph

Data over the Internet is shared using the *Web* which is basically a distributed logical network of *Web pages*. A node (page) in the Web has a domain name which is a string of literals separated by periods. The main protocol for data transfer over the Web is the *Hyper Text Transfer Protocol* (HTTP) which has various commands, for example CONNECT to connect to the Web page and GET which requests a representation of the required resource. A typical Web page has a name ending with "html" which means the page is written using the Hyper Text Markup Language.

In the common usage, a Web page points to another Web page. Thus, the Web may be conveniently represented by a digraph as shown in Fig. 15.6 where Üsküdar University is pointed by universities Web page. The university page has a link to the Faculty of Engineering and Natural Sciences which is also pointed by the "Faculties"

15.2.2   Page Rank Algorithm

The importance or the *rank* of a Web page may be associated with the number of pages that reference it. This score for a Web page may be used to display the more important pages on top when a search is made. Page rank is like a fluid that runs through the nodes of the network accumulating at important nodes.

The page rank algorithm (PRA) allow calculation of page rank values based on the pages that reference it. Each page is assigned a rank value of $1/n$ in the Web graph initially where n is the number of Web pages. Thereafter, a rank value of a page is distributed to its outgoing edges evenly and the rank value of a page is calculated as the sum of the weights of its ingoing edges. The general idea is that the rank of a page increases with the number and weights of edges that point to it. The pseudocode of this algorithm is displayed in Algorithm 15.1 and it can be shown the page rank values converge as $k \rightarrow \infty$

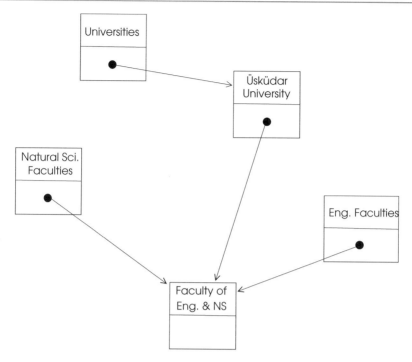

Fig. 15.6 A part of Web Graph

Algorithm 15.1 *Page rank algorithm*

1: **Input**: $P = \{p_1, ..., p_n\}$ ▷ Web pages
2: **Output**: $R = \{rank_1, ..., rank_n\}$ ▷ Page rank values
3: $E_p(in) \leftarrow$ ingoing edges to page p
4: $E_p(out) \leftarrow$ outgoing edges from page p
5: **for all** $p \in P$ **do**
6:     $rank_p \leftarrow 1/n$
7: **end for**
8: **for** $x = 1$ to k **do**
9:     **for all** $p \in P$ **do**
10:         **for all** $e \in E_p(out)$ **do**
11:             $w_e \leftarrow rank_p/|E_p(out)|$
12:         **end for**
13:         $rank_p \leftarrow \sum_{e \in E_p(in)} w_e$
14:     **end for**
15: **end for**

Fig. 15.7 A sample Web subgraph

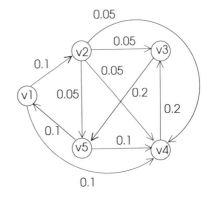

Fig. 15.8 Calculation of rank values for the sample Web subgraph

	v_1	v_2	v_3	v_4	v_5
n_out	2	3	1	1	2
$k=1$ out_vals	0.1	0.05	0.2	0.2	0.1
$rank(v_i)$	0.1	0.1	0.2	0.5	0.25
$k=2$ out_vals	0.05	0.03	0.2	0.5	0.13
$rank(v_i)$	0.25	0.05	0.03	1	0.3

Figure 15.7 displays a small Web subgraph with initial edge weights, and the calculation of rank values for Web pages using the Page Rank algorithm is shown in Fig. 15.8. Initially, each page starts with $1/n = 1/5 = 0.2$ value and this value is distributed evenly to all outgoing edges of each page. For example, page v_5 has two outgoing edges, thus, the weights for these pages are $0.2/2 = 0.1$ as shown in the first line of the table for the fist iteration when $k = 1$. The second line shows the page rank values obtained by summing the weights of input edge weights to a page and the same process is repeated for $k = 2$. We can see page v_4 has the highest rank as can be seen by 4 pages referencing it. A page that does not point to any page may have a high rank value and this can be corrected by implementing damping factor d that reduces page rank values by $(1 - d)/n$ [10].

15.3 Ad hoc Wireless Networks

A wireless network comprises computational nodes that communicate using wireless communication channels. An *ad hoc* wireless network does not have a fixed communication structure in contrast to an infrastructured network which typically provides a wired basic communication backbone to wireless hosts. Ad hoc wireless networks do not have a static communication structure and for this reason, they use *multi-hop communication* to transfer messages in which a message goes through a number of nodes each broadcasting it until the message reaches the destination.

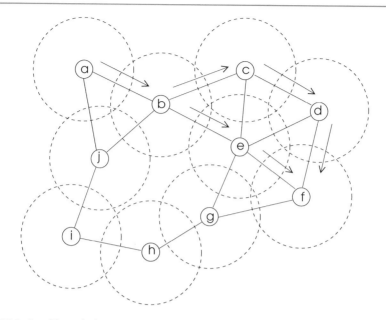

Fig. 15.9 An ad hoc wireless network

Two types of ad hoc networks have found common usage; *mobile ad hoc networks* (MANETs) and wireless sensor networks (WSNs). Computation structure used in a rescue operation or military operation is a typical MANET application. A wireless sensor network (WSN) consists of small computational nodes equipped with antennas and batteries, and communicate using radio waves. These networks find numerous applications ranging from intelligent farming to e-health to intelligent buildings to border surveillance systems.

Whether a MANET or a WSN, an ad hoc network may be represented as a graph and various graph properties, concepts and proven results of graph theory become readily available to be implemented in these networks. An ad hoc wireless network is depicted in Fig. 15.9 with the transmission range of a node shown in dashed circles. If two nodes are within transmission ranges of each other, we can connect them with an edge assuming communication is bidirectional. For example, a message from node a to f can be sent through the path a, b, e, f or path a, b, c, d, f using multi-hop communication.

15.3.1 Routing in ad hoc Networks

We do not have a central control in various computer networks, instead, each node of the network acts independently. The nodes cooperate to finish an overall task in such a *distributed system* and algorithms performed in such systems are called *distributed algorithms*.

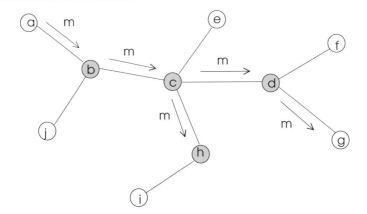

Fig. 15.10 Routing using MCDS in an ad hoc network

A dominating set (DS) of a graph $G = (V, E)$ was defined in Sect. 13.4 as a subset V' of V such that any vertex v in V is either a member of V' or adjacent to a vertex in V'. A connected DS (CDS) has a path between each member vertex pairs that go through CDS vertices only. A minimal CDS (MCDS) may be used to form a communication backbone in a wireless ad hoc network as noted as we will briefly re-review the needed procedure. Let us consider the ad hoc wireless network in Fig. 15.10 where MCDS nodes are shown in grey. A node a that wants to send a message m simply relays the message to its dominating node b for example. The node b broadcasts m over the backbone and any node that receives m checks whether destination is one of the nodes it dominates in which case it simply delivers the message to its destination node g only as in this example. Otherwise m is broadcast over the backbone by the MCDS nodes until it reaches its destination. This way, message is transmitted mainly over the backbone thereby reducing the traffic.

We will describe a distributed algorithm, which is executed by all nodes in the network, due to Wu and Lin to form a MCDS in a wireless ad hoc network. A node assigned to MCDS is assigned the black color and dominated nodes are white at the end of the algorithm. This algorithm performs the following steps in the first phase.

1. Each node exchanges the identifiers of its neighbors with all of its neighbors.
2. Any node that finds it has two unconnected neighbors marks itself to be in MCDS by changing its color to black.
3. Each node sends its color to its neighbors.

The second phase called *pruning* is needed to remove redundant nodes from the MCDS with the following rules.

1. if a node u finds all of its neighbors are covered with a neighbor v that has a higher identifier than u, u removes itself by changing its color back to white.

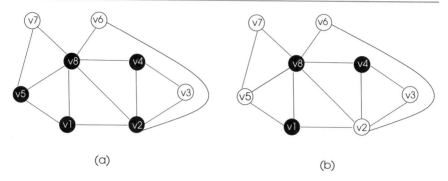

(a)   (b)

Fig. 15.11 Running of Wu's algorithm in a sample graph

2. If a node u finds that the union of the neighbors of its two MCDS neighbors v and w cover all of its neighbors and the identifiers of v and w are both greater than that of u, u removes itself from the MCDS by changing its color to white.

Running of this algorithm in a small network is depicted in Fig. 15.11. In the first phase, nodes v_1, v_2, v_4, v_5 and v_8 assign black color to themselves as they have at least two unconnected neighbors shown in (a). Then in the second phase, Node v_5 changes its color back to white by the first rule as a higher identifier v_8 covers both of its neighbors. Node v_2 also changes its color to white by the second rule as the union of its higher identifier neighbors v_4 and v_8 cover all of its neighbors which result in the MCDS in (b).

15.3.2 Clustering and Spanning Tree Construction in a WSN

A WSN commonly employs a spanning tree structure to communicate with the sink node. Sensors collect data and each node sends its data and data received from its children to its parent in a *convergecast* operation. The sink node sends a data typically in command form to all nodes using *broadcast* operation where each node sends received message to its children. A simple command would initiate taking sensor measurements. These operations are depicted in Fig. 15.12 with $d(a)$ showing the data sensor a obtains for example. The data collected from children may the temperature sensed in the sensor environment and a node may send the average value of the data received from its children and data itself measured to its parent. This process is called *data aggregation* where data is sent upwards in a compressed form to the sink.

On the other hand, clustering in a WSN is practical as there is a large number of nodes commonly exceeding hundreds, and communication using clusterheads is preferred rather than sending messages among many nodes. An algorithm due to Erciyes et al. performs these two tasks; namely, spanning tree construction and

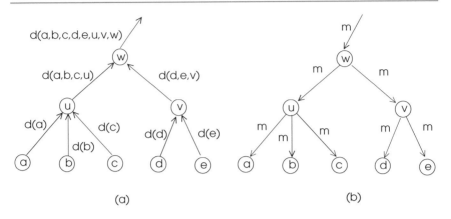

Fig. 15.12 a Convergecast **b** Broadcast operation in a WSN

clustering in a WSN simultaneously where each node has a unique identifier [5]. The steps of this algorithm is as follows.

1. The sink node initiates the algorithm by broadcasting a *probe* message which contains its identifier and a field called *hop_count* with 0 value to all its neighbors.
2. Any node v that receives *probe* message does the following.

   a. If probe is received for the first time, it marks the sender u as its parent, increments the *hop_count* value, inserts its identifier in the message and broadcasts the message. It also sends *ack* message to its parent u with its identifier.
   b. When the probe is received for the first time, if the *hop_count* in message is less than or equal the cluster *hop_limit* value, it includes itself in the cluster by assigning the initiator u as its clusterhead.
   c. If probe was received before, it sends negative acknowledgement message *nack* to the sender.
   d. A timeout after a node u sends a *probe* message means u has its parent as the only neighbor, that is, u is a leaf node.

3. A node u that receives an *ack* message from node v stores v as its child.

The working of this algorithm is depicted in Fig. 15.13 with clusterheads shown in black and *hop_limit* $= 2$. The time complexity of this algorithm is $O(d)$ where d is the diameter of the network and it requires $O(n)$ messages [5].

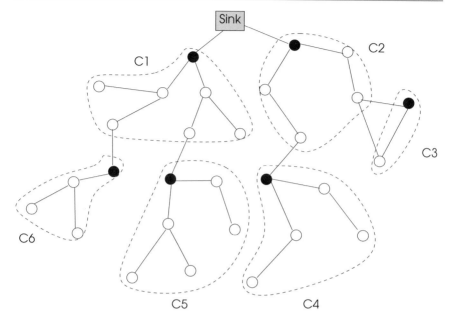

Fig. 15.13 Spanning tree formation and clustering in a WSN

15.4 Biological Networks

Biological networks are large and complex, for example, protein-protein interaction (PPI) networks have proteins as nodes and signalling among proteins are the edges. A protein is the basic component of a cell and carries various tasks within the cell. Understanding the structure of a PPI network may provide insight to understanding health and disease states of an organism [3]. A phylogenetic tree displays hereditary relationships among individuals. We will look at two distinct problems in biological networks; network motif finding and network alignment.

15.4.1 Network Motifs

A network motif is a recurring subgraph structure in a network. Finding the statistical significance of a network motif may show the structure and building blocks of an organism since a frequently occurring motif is assumed not to occur by chance and should have some significance. Whenever such a network motif is detected, our next step would be investigating the functionality of such a motif. Various commonly found motifs in biological networks are depicted in Fig. 15.14.

Feed-forward loop is commonly found in many gene systems. Detecting of this motif in an example graph is shown in Fig. 15.15 where subgraphs M_1, M_2, M_3 and

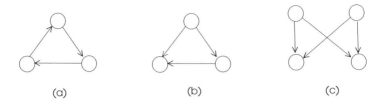

Fig. 15.14 Motifs in biological networks, **a** Feedback loop **b** Feed-forward loop **c** Bi-fan

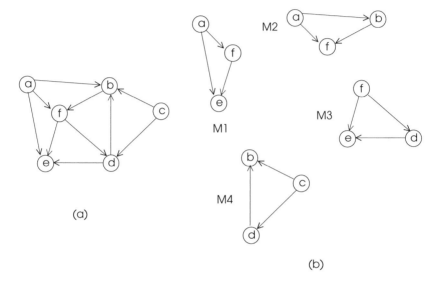

Fig. 15.15 a A sample graph **b** Feed-forward loops this graph

M_4 with this motif are detected. Finding motifs is a difficult problem and heuristics are commonly used.

15.4.2 Network Alignment

When a number of graphs are used to represent certain biological networks, it is of interest to find the similarity of these networks. The graph alignment or more commonly, the *network alignment* problem is to discover the similarities between networks and thus, to deduce genetic relations between them. Let $G = (V_1, E_1)$ and $H = (V_2, E_2)$ be two graphs we need to investigate, and we have a function $f : V_1 \rightarrow V_2$ that maps vertices of G to the vertices of H. The quality of alignment is given by the function $Q(G, H, f)$ and the main target of any alignment method is to maximize Q. Network alignment can be performed as follows [3]

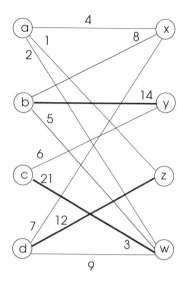

Fig. 15.16 Weighted bipartite graph matching for network alignment example

- *Local or Global alignment*: Local alignment is the process of finding similarities between small subgraphs of two or more networks. Global alignment aims to find similarities over the whole graphs representing the biological networks.
- *Pairwise or Multiple alignment*: The pairwise alignment methods input two graphs and the latter inputs more than two graphs.
- *Node or topological similarity based alignment*: Certain attributers of vertices may be considered in node similarity-based alignment methods and the latter refers to topological similarity. A combination of both methods may be used by giving different weights to each method [3].

Network alignment problem may be formed as a weighted bipartite graph matching problem where the edges between the two vertex partitions have weight dependent on the similarity of vertices. Our aim is then to form the maximal weighted matching and thus, the simple greedy algorithm described in Sect. 13.2.3 with $O(m^2)$ time complexity may be used. Two simple networks aligned using this method is shown in Fig. 15.16 where node b is aligned to node y, node c is aligned to node w and node d is aligned to node z and only nodes a and x are not aligned due to their low resemblances to other nodes.

15.5 Social Networks

A social network has a number of persons who have some kind of relationship such as the students of a class. A social network may be represented by a graph with nodes showing the persons or groups and edges their interaction. A simple social network

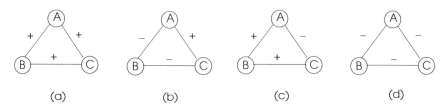

Fig. 15.17 Relationships among three persons

is the friendship network with edges displaying acquaintances between persons for example. We will define some concepts that will be helpful in the analysis of social networks in the following.

15.5.1 Relationships

We can label a relationship between two persons as positive to mean they are friends or negative to mean they dislike each other. Considering three persons A, B and C; four possible relationships between them is depicted in Fig. 15.17 [4]. A *balanced* relationship is defined as a stable relationship and we have two balanced states in these four states. For example, the relationship is stable when three are all friends as in (a). When two persons A and B are friends but both dislike person C, we again have a stable state. The other two states are not balanced; for example, a person A likes B and C but B and C dislike each other. Lastly, when all persons dislike each other, this is again stable since they will be reluctant to form a relationship anyway. We assumed like/dislike property is bidirectional, however, a person A may like another person B but B may dislike A in reality. A digraph may represent a social network more conveniently in such a case.

15.5.2 Structural Balance

Let a graph $G = (V, E)$ represent a social network. Then there are two cases for the social network represented by G to be balanced; either all nodes like each other or vertices of G can be divided into two distinct groups A and B with the following properties.

- All nodes in A like each other.
- All nodes in B like each other.
- Each node in A dislikes every node in B.
- Each node in B dislikes every node in A.

This property is proposed in the *balance theorem* by Harary [9]. Consider the graph of Fig. 15.18 which represents a social network, we have group $A = \{b, c, f, g\}$ and group $B = \{a, d, h\}$ obeying the balance theorem, therefore this social network is

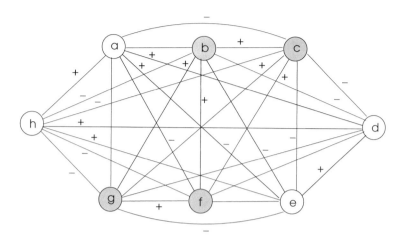

Fig. 15.18 A balanced social network

balanced. This situation was considered as one of the reasons why the world wars lasted long, every nation in one group was enemy with every nation in the other group, and every nation in one group was friends with all other nations in its group; thus, the war network was stable.

It is of interest to detect closely interacting persons/groups in a social network for a number of reasons. This process of finding densely communicating subnetworks in a social network is termed *community detection* which is basically clustering method reviewed in Sect. 15.1.2.

15.6 Review Questions

1. What is clustering coefficient of a vertex in a graph and why is it important?
2. What is matching index and what does it provide?
3. State the types of centralities in a large graph.
4. What are the main types of large networks?
5. What is a complex network and what are the two main characteristics of complex networks?
6. What is the main idea of the Page Rank algorithm?
7. How may routing be performed in an ad hoc network?
8. Why building a spanning tree is needed in a WSN?
9. What is network motif search in a biological network and why is it important?
10. What is network alignment problem in a biological network and what is its significance?
11. What are the balanced configurations of a three people friendship network?
12. What is structural balance in a social network?
13. State the steps of the MST-based clustering algorithm in a social network.

15.7 Chapter Notes

Graphs are used to represent various networks including the Internet, the Web, ad hoc wireless networks, biological networks and social networks. We reviewed basic principles of employing graphs for these networks. Having a graph model provides various algorithms such as minimum spanning tree formation, breadth-first-search, depth-first-search to our use and many problems in these networks may have optimal or suboptimal solutions using basic graph algorithms.

All of these networks are large and they have properties such as small-world and scale-free which are not found in a random network. We started this chapter by the analysis of large networks and showed how various problems in these networks may have a suboptimal solution using graph properties. For example, building a minimal connected dominating set of an ad hoc network provides a backbone for efficient message transfer. In another example, we saw how network alignment problem in biological networks can be solved to some extent by a graph theoretic bipartite matching algorithm. Clustering is a well-studied topic in various disciplines including Computer Science, Statistics and Bioinformatics and we can make use of some graph theoretical property while clustering. Graph applications are numerous and have a much wider spectrum than discussed in this chapter, ranging from machine theory to bioinformatics and to stock exchange. It is widely believed that graphs will have increasingly many more applications in very diverse fields of science.

Exercises

1. Plot the degree distribution of the graph of Fig. 15.19.
2. Calculate clustering coefficients of the vertices in Fig. 15.20.
3. Calculate matching indices for the vertex pairs in Fig. 15.20.
4. Work out all page rank values for three iterations of the Web graph shown in Fig. 15.21.
5. Find all feedback loop motifs in the digraph of Fig. 15.22.
6. Two biological networks $N_1 = \{a, b, c, d, e\}$ and $N_2 = \{x, y, z, w, t\}$ with 5 elements each are are to be aligned. Their affinity matrix A with each element $A[i, j]$ showing the closeness score of element i of N_1 to element j of N_2 has the following contents with ∞ meaning no connection.

$$
\begin{array}{c}
 & \begin{array}{ccccc} x & y & z & w & t \end{array} \\
A = \begin{array}{c} a \\ b \\ c \\ d \\ e \end{array} & \left(\begin{array}{ccccc}
6 & \infty & 22 & \infty & 17 \\
7 & 3 & \infty & \infty & 24 \\
5 & \infty & \infty & \infty & 2 \\
20 & \infty & \infty & 11 & 4 \\
\infty & \infty & 1 & 16 & 5
\end{array} \right)
\end{array}
$$

Work out the alignment using the simple greedy algorithm.

Fig. 15.19 Example graph for Exercise 1

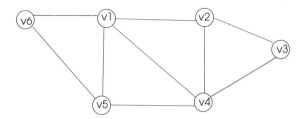

Fig. 15.20 Sample graph for Exercises 2 and 3

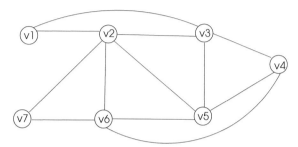

Fig. 15.21 Web graph for Exercise 4

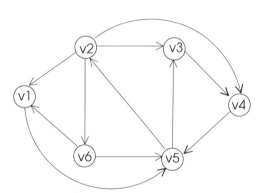

Fig. 15.22 Sample graph for Exercise 5

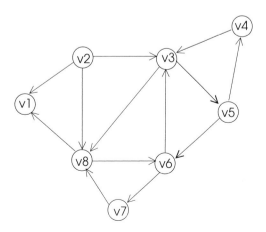

References

1. Barabasi AL, Albert R (1999) Emergence of scaling in random networks. Science 286:509–512
2. Brelaz D (1979) New methods to color the vertices of a graph. Commun ACM 22(4):
3. Erciyes K (2015) Distributed and sequential algorithms for bioinformatics. Springer, Berlin
4. Erciyes K (2014) Complex networks, an algorithmic perspective. CRC Press, Boca Raton
5. Erciyes K, Ozsoyeller D, Dagdeviren O, (2008) Distributed algorithms to form cluster based spanning trees in wireless sensor networks. ICCS, (2008) LNCS. Springer, Berlin, pp 519–528
6. Erdos P, Renyi A (1959) On random graphs, i. Publicationes Mathematicae (Debrecen) 6:290–297
7. Garey MR, Johnson DS (1979) Computers and intractability. Freeman, New York, W.H
8. Grimmet GR, McDiarmid CJH (1975) On coloring random graphs. Mathematical proceedings of the Cambridge philosophical society 77:313–324
9. Harary F (1953) On the notion of balance of a signed graph. Michigan Math J 2(2):143–146
10. Kleinberg J (1999) Authoritative sources in hyperlinked environment. J ACM 46(5):604–632
11. Milgram S (1967) The small world problem. Psychol Today 2:60–67

Pseudocode Conventions

A

A.1 Introduction

In this part, the pseudocode conventions for writing an algorithm is presented. The conventions we use follow the modern programming guidelines and are similar to the ones used in [1,2]. Every algorithm has a name specified in its heading and each line of an algorithm is numbered to provide citation. The first part of an algorithm usually starts by its inputs. Blocks within the algorithm are shown by indentations. The pseudocode conventions adopted are described as data structures, control structures and distributed algorithm structure as follows.

A.2 Data Structures

Expressions are built using constants, variables and operators as in any functional programming language and yield a definite value. *Statements* are composed of expressions and are the main unit of executions. All statements are in the form of numbered lines. Declaring a variable is done as in languages like Pascal and C where its type precedes its label with possible initialization as follows:

$$\textbf{set of int} \quad neighbors \leftarrow \{\emptyset\}$$

Here we declare a set called *neighbors* of a vertex in a graph each element of which is an integer. This set is initialized to $\{\emptyset\}$ (empty) value. The other commonly used variable types in the algorithms are *boolean* for boolean variables and *message types* for the possible types of messages. For assignment, we use $\leftarrow$ operator which shows that the value on the right is assigned to the variable in the left. For example, the statement:

$$a \leftarrow a + 1$$

© Springer Nature Switzerland AG 2021
K. Erciyes, *Discrete Mathematics and Graph Theory*, Undergraduate Topics
in Computer Science, https://doi.org/10.1007/978-3-030-61115-6_A

Table A.1 General algorithm conventions

Notation	Meaning
$x \leftarrow y$	Assignment
$=$	Comparison of equality
$\neq$	Comparison of inequality
true, false	Logical true and false
null	Non-existence
$\triangleright$	Comment

Table A.2 Arithmetic and logical operators

Notation	Meaning
$\neg$	Logical negation
$\wedge$	Logical and
$\vee$	Logical or
$\oplus$	Logical exclusive-or
x/y	x Divided by y
$x.y$ or xy	Multiplication

increments the value of the integer variable a. Two or more statements in a line are separated by semicolons and comments are shown by $\triangleright$ symbol at the end of the line as follows:

$$1 : a \leftarrow 1; c \leftarrow a + 2; \qquad \triangleright \quad c \; is \; now \; 3$$

General algorithmic conventions are outlined in Table A.1, and Table A.2 summarizes the arithmetic and logical operators used in the text with their meanings.

Sets instead of arrays are frequently used to represent a collection of similar variables. Inclusion of an element u to a set S can be done as follows:

$$S \leftarrow S \cup \{u\}$$

and deletion of an element v from S is performed as follows:

$$S \leftarrow S \setminus \{v\}$$

Table A.3 shows the set operations used in the text with their meanings.

A.3 Control Structures

In the sequential operation, statements are executed consecutively. Branching to another statement can be done by *selection* described below.

Table A.3 Set operations

Notation	Meaning
$\lvert S \rvert$	Cardinality of S
$\emptyset$	Empty set
$u \in S$	u is a member of S
$S \bigcup R$	Union of S and R
$S \bigcap R$	Intersection of S and R
$S \setminus R$	Set subtraction
$S \subset R$	S is a proper subset of R
$max/min\ S$	Maximum/minimum value of the elements of S
$max/min\{....\}\ S$	Maximum/minimum value of a collection of values

Selection

Selection is performed using conditional statements which are implemented using *if-then-else* in the usual fashion and indentation is used to specify the blocks as shown in the example code segment below:

Algorithm A.1 *if-then-else structure*

1: **if** *condition* **then**   ▷ first check
2:      *statement*1
3:      **if** *condition2* **then**   ▷ second (nested) *if*
4:          *statement2*
5:      **end if**   ▷ end of second *if*
6: **else if** *condition3* **then**   ▷ *else if* of first *if*
7:      *statement3*
8: **else**
9:      *statement4*
10: **end if**   ▷ end of first *if*

In order to select from a number of branches, *case-of* construct is used. The expression within this construct should return a value which is checked against a number of constant values and the matching branch is taken as follows:

1. **case** *expression* **of**
2. $constant_1: statement_1$
3. $\vdots$
4. $constant_n: statement_n$
5. **end case**

Repetition

The main loops in accordance with the usual high level language syntax are the *for*, *while* and *loop* constructs. The *for-do* loop is used when the count of iterations can be evaluated before entering the loop as follows:

1. **for** $i \leftarrow 1$ **to** n **do**
2. $\vdots$
3. **end for**

The second form of this construct is the *for all* loop which arbitrarily selects an element from the set specified and iterates until all members of the set are processed as shown below where a set S with three elements and an empty set R are given and each element of S is copied to R iteratively.

1. $S \leftarrow \{3, 1, 5\}; R \leftarrow \emptyset$
2. **for all** $u \in S$ **do**
3. $R \leftarrow R \bigcup \{u\}$
4. **end for**

For the indefinite cases where the loop may not be entered at all, the *while-do* construct may be used where the boolean expression is evaluated and the loop is entered if this value is true as follows:

1. **while** *boolean expression* **do**
2.   *statement*
3. **end for**

References

1. Cormen TH, Leiserson CE, Rivest RL, Stein C (2001) Introduction to algorithms. MIT Press, Cambridge
2. Smed J, Hakonen H (2006) Algorithms and networking for computer games. Wiley Ltd, New Jersey. ISBN 0-470-01812-7

Index

© Springer Nature Switzerland AG 2021
K. Erciyes, *Discrete Mathematics and Graph Theory*, Undergraduate Topics
in Computer Science, https://doi.org/10.1007/978-3-030-61115-6

Printed in the United States
By Bookmasters